NATURAL DISASTER RESEARCH, PREDICTION AND MITIGATION

NATURAL DISASTERS

RESPONSE, RECOVERY AND ASSISTANCE

Natural Disaster Research, Prediction and Mitigation

Additional books and e-books in this series can be found on Nova's website under the Series tab.

NATURAL DISASTER RESEARCH, PREDICTION AND MITIGATION

NATURAL DISASTERS

RESPONSE, RECOVERY AND ASSISTANCE

SONJA TORRES
EDITOR

snova
New York

Additional color graphics may be available in the e-book version of this book.

Library of Congress Cataloging-in-Publication Data

ISBN: 978-1-53616-407-7

Published by Nova Science Publishers, Inc. † New York

CONTENTS

PREFACE

In 2017, four sequential disasters—hurricanes Harvey, Irma, Maria, and the California wildfires—created an unprecedented demand for federal disaster response and recovery resources. Chapter 1 provides observations on federal and state preparedness and response coordination for hurricanes Harvey and Irma in Texas and Florida, and the California wildfires; federal preparedness for and response to hurricanes Irma and Maria in Puerto Rico and the U.S. Virgin Islands; and existing and emerging disaster recovery challenges highlighted by these disasters. Chapter 2 provides information on federal financial assistance provided to the Gulf States after major disasters were declared in Alabama, Florida, Louisiana, Mississippi, and Texas in response to the widespread destruction that resulted from Hurricanes Katrina, Rita, and Wilma in 2005 and Hurricanes Gustav and Ike in 2008.

Chapter 1 - In 2017, four sequential disasters—hurricanes Harvey, Irma, Maria, and the California wildfires—created an unprecedented demand for federal disaster response and recovery resources. According to FEMA, 2017 included three of the top five costliest hurricanes on record. The National Oceanic and Atmospheric Administration estimated that the cumulative damages from weather and climate related disasters in the United States were over $300 billion in 2017 alone. As of June 2018, Congress had appropriated over $120 billion in supplemental funding for response and recovery related to the 2017 hurricanes and wildfires. Further, in October 2017, close to 14,000 federal employees were deployed in response to the disasters. Given the scale and cost of these disasters, Congress and others have raised questions about the federal response and various recovery challenges that have arisen since the disasters. This chapter provides GAO's observations on: (1) federal and state preparedness and response coordination for hurricanes Harvey and Irma in Texas and Florida, and the California wildfires; (2) federal preparedness for and response to hurricanes Irma and Maria in Puerto Rico and the U.S. Virgin Islands; and (3) existing and emerging disaster recovery challenges highlighted by these disasters. GAO analyzed FEMA policies, procedures, guidance, and data specific to disaster response and recovery programs. GAO focused on the busiest period of disaster response activity for the federal government— August 2017 through January 2018, with select updates on recovery efforts and obtained updates through June 2018. In October and November 2017, GAO teams made site visits to hurricane damaged areas in Texas, Florida, Puerto Rico, and the U.S. Virgin Islands. At these locations, GAO visited FEMA joint field operation locations and interviewed FEMA, Department of Defense, and other federal officials about response and recovery operations, visited disaster recovery centers, and observed damage. GAO also interviewed FEMA officials responsible for wildfire response and recovery efforts in California. Additionally, GAO interviewed state and territorial emergency management officials or their designee in Texas, Florida, California, Puerto Rico, and the U.S. Virgin Islands, as well as officials from eight cities and counties in Texas, Florida, and California (selected based on their proximity to the disaster impacted areas and their availability) to discuss their observations on the federal response in their respective jurisdictions. While the perspectives of these officials are not generalizable, they provided valuable insights into the federal response to the 2017 disasters. This chapter includes 10 appendices that provide further details and data on federal response and recovery efforts. These areas cover key issues and challenges that GAO believes are critical for assessing the federal response and warrant continued Congressional and agency oversight during disaster recovery. GAO is not making recommendations in this chapter, but has ongoing work that will address various response and recovery programs and challenges in more detail. GAO will make recommendations, as appropriate, once this work is completed. In commenting on a draft of this chapter, DHS stated that the report highlighted the challenges of the complicated response and recovery efforts as well as provided insights into these efforts. DHS also noted that it is continuing to apply lessons learned from 2017 to improve its future program delivery and response efforts.

Chapter 2 - This chapter provides information on federal financial assistance provided to the Gulf States after major disasters were declared in Alabama, Florida, Louisiana, Mississippi, and Texas in response to the widespread destruction that resulted from Hurricanes Katrina, Rita, and Wilma in 2005 and Hurricanes Gustav and Ike in 2008. Though the storms happened over a decade ago, Congress has remained interested in the types and amounts of federal assistance that were provided to the Gulf Coast for several reasons. This includes how the money has been spent, what resources have been provided to the region, and whether the money has reached the intended people and entities. The financial information is also useful for congressional oversight of the federal programs provided in response to the storms. It gives Congress a general idea of the federal assets that are needed and can be brought to bear when catastrophic disasters take place in the United States. Finally, the financial information from the storms can help frame the congressional debate concerning federal assistance for current and future disasters. The financial information for the 2005 and 2008 Gulf Coast storms is provided in two sections of this chapter:

- Table 1 of Section I summarizes disaster assistance supplemental appropriations enacted into public law primarily for the needs associated with the five hurricanes, with the information categorized by federal department and agency; and

- Section II contains information on the federal assistance provided to the five Gulf Coast states through the most significant federal programs, or categories of programs.

The financial findings in this chapter include:

- Congress has appropriated roughly $121.7 billion in hurricane relief for the 2005 and 2008 hurricanes in 10 supplemental appropriations statutes.
- The appropriated funds have been distributed among 11 departments, 3 independent agencies/entities, numerous sub-entities, and the federal judiciary.
- Congress appropriated almost half of the funds ($53.8 billion, or 44% of the total) to the Department of Homeland Security, most of which went to the Disaster Relief Fund (DRF) administered by the Federal Emergency Management Agency (FEMA).
- Congress targeted roughly 22% of the total appropriations (almost $27 billion) to the Department of Housing and Urban Development for community development and housing programs.
- Approximately 20% ($25 billion) was appropriated to Department of Defense entities: $15.6 billion for civil construction and engineering activities undertaken by the Army Corps of Engineers and $9.2 billion for military personnel, operations, and construction costs.
- FEMA has reported that roughly $5.9 billion has been obligated from the DRF after Hurricanes Katrina, Rita, and Wilma to save lives and property through mission assignments made to over 50 federal entities and the American Red Cross, $160.4 million after Hurricane Gustav through 32 federal entities, and $441 million after Hurricane Ike through 30 federal entities. In total, federal agencies obligated roughly $6.5 billion for mission assignments after the five hurricanes.

- The Small Business Administration approved almost 177,000 applications in the region for business, home, and economic injury loans, with a total loan value of almost $12 billion.
- The Department of Education obligated roughly $1.8 billion to the five states for elementary, secondary, and higher education assistance.

This chapter also includes a brief summary of each hurricane and a discussion concerning federal to state cost-shares. Federal assistance to states is triggered when the President issues a major disaster declaration. In general, once declared the federal share for disaster recovery is 75% while the state pays for 25% of recovery costs. However, in some cases the federal share can be adjusted upward when a sufficient amount of damage has occurred, or when altered by Congress (or both). In addition, how much federal assistance is provided to states for major disasters is influenced not only by the declaration, but also by the percentage the federal government pays for the assistance. This chapter includes a cost-share discussion because some of these incidents received adjusted cost-shares in certain areas.

In: Natural Disasters
Editor: Sonja Torres
ISBN: 978-1-53616-407-7

Chapter 1

2017 Hurricanes and Wildfires: Initial Observations on the Federal Response and Key Recovery Challenges*

United States Government Accountability Office

Abbreviations

DHS	Department of Homeland Security
DRF	Disaster Relief Fund
DOD	Department of Defense
EMAC	Emergency Management Assistance Compact
ESFs	emergency support functions
FCO	Federal Coordinating Officer
FEMA	Federal Emergency Management Agency

* This is an edited, reformatted and augmented version of United States Government Accountability Office; Report to Congressional Addressees, Publication No. GAO-18-472, dated September 2018.

FQS	FEMA Qualification System
IA	Individual Assistance
IMAT	Incident Management Assistance Team
IT	information technology
JFO	joint field office
PA	Public Assistance
PREPA	Puerto Rico Electric Power Authority
PROMESA	Puerto Rico Oversight, Management, and Economic
SRIA	Sandy Recovery Improvement Act of 2013
USACE	U.S. Army Corps of Engineers

Why GAO Did This Study

In 2017, four sequential disasters—hurricanes Harvey, Irma, Maria, and the California wildfires—created an unprecedented demand for federal disaster response and recovery resources. According to FEMA, 2017 included three of the top five costliest hurricanes on record.

The National Oceanic and Atmospheric Administration estimated that the cumulative damages from weather and climate related disasters in the United States were over $300 billion in 2017 alone. As of June 2018, Congress had appropriated over $120 billion in supplemental funding for response and recovery related to the 2017 hurricanes and wildfires. Further, in October 2017, close to 14,000 federal employees were deployed in response to the disasters.

Given the scale and cost of these disasters, Congress and others have raised questions about the federal response and various recovery challenges that have arisen since the disasters. This chapter provides GAO's observations on: (1) federal and state preparedness and response coordination for hurricanes Harvey and Irma in Texas and Florida, and the California wildfires; (2) federal preparedness for and response to hurricanes Irma and Maria in Puerto Rico and the U.S. Virgin Islands; and

(3) existing and emerging disaster recovery challenges highlighted by these disasters.

GAO analyzed FEMA policies, procedures, guidance, and data specific to disaster response and recovery programs. GAO focused on the busiest period of disaster response activity for the federal government— August 2017 through January 2018, with select updates on recovery efforts and obtained updates through June 2018. In October and November 2017, GAO teams made site visits to hurricane damaged areas in Texas, Florida, Puerto Rico, and the U.S. Virgin Islands. At these locations, GAO visited FEMA joint field operation locations and interviewed FEMA, Department of Defense, and other federal officials about response and recovery operations, visited disaster recovery centers, and observed damage. GAO also interviewed FEMA officials responsible for wildfire response and recovery efforts in California.

Additionally, GAO interviewed state and territorial emergency management officials or their designee in Texas, Florida, California, Puerto Rico, and the U.S. Virgin Islands, as well as officials from eight cities and counties in Texas, Florida, and California (selected based on their proximity to the disaster impacted areas and their availability) to discuss their observations on the federal response in their respective jurisdictions. While the perspectives of these officials are not generalizable, they provided valuable insights into the federal response to the 2017 disasters.

This chapter includes 10 appendices that provide further details and data on federal response and recovery efforts. These areas cover key issues and challenges that GAO believes are critical for assessing the federal response and warrant continued Congressional and agency oversight during disaster recovery.

GAO is not making recommendations in this chapter, but has ongoing work that will address various response and recovery programs and challenges in more detail. GAO will make recommendations, as appropriate, once this work is completed.

In commenting on a draft of this chapter, DHS stated that the report highlighted the challenges of the complicated response and recovery efforts as well as provided insights into these efforts. DHS also noted that it is continuing to apply lessons learned from 2017 to improve its future program delivery and response efforts.

WHAT GAO FOUND

Federal and state preparedness and coordination efforts prior to and after the 2017 hurricane and wildfire disasters facilitated the response in Texas, Florida, and California. Specifically, the Federal Emergency Management Agency (FEMA) and state emergency management officials implemented various preparedness actions prior to landfall of the hurricanes and during the wildfires— such as predeploying federal personnel to support response efforts; colocating federal, state, and local emergency managers; and pre-staging and delivery of commodities like food and water. Further, according to FEMA and state officials, preexisting coordination mechanisms and relationships also facilitated response efforts in each state. For example, FEMA and each state had conducted numerous emergency exercises in the years prior to these disasters and had developed relationships during response to prior disasters that led to accelerated decision-making during the 2017 disasters. Federal and state officials emphasized that there were certainly unprecedented challenges during these disasters—such as deploying a sufficient and adequately-trained FEMA disaster workforce—and lessons learned, but prior response coordination efforts helped to quickly and effectively resolve many of these challenges. The federal government provided significant support to Puerto Rico and the U.S. Virgin Islands in response to Hurricanes Irma and Maria, but faced numerous challenges that complicated response efforts. FEMA efforts in Puerto Rico alone were the largest and longest single response in the agency's history. As of April 2018, FEMA had obligated over $12 billion for response and recovery for Hurricane Maria (see Figure below) reflecting the scale and complexity of efforts given the widespread damage. FEMA tasked federal agencies with over 1,000 response mission assignments for Hurricanes Maria and Irma in the territories at a cost of over $5 billion, compared to about 400 such assignments for Hurricanes Harvey and Irma and the California wildfires combined. For example, FEMA assigned the U.S. Army Corps of Engineers the mission to install over 1,700 emergency electricity generators in Puerto Rico, compared to the 310 for the response to Hurricane Katrina.

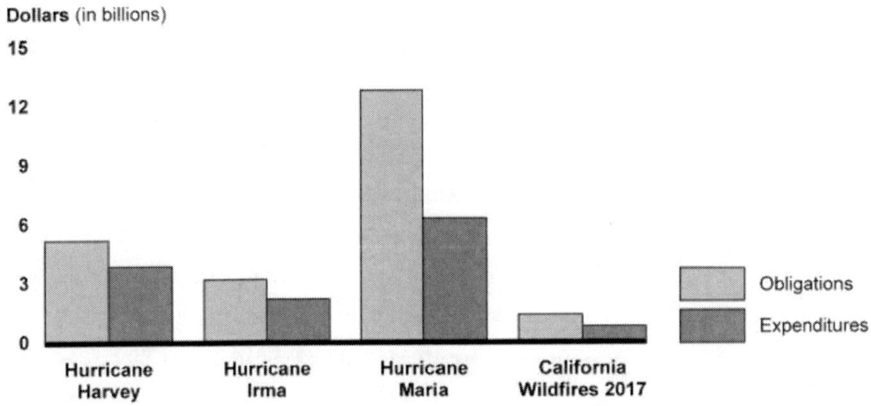

Source: GAO analysis of Federal Emergency Management Agency data. | GAO-18-472.

Note: An obligation is a definite commitment that creates a legal liability of the government for the payment of goods and services ordered or received. An expenditure is an amount paid by federal agencies by cash or cash equivalent, during the fiscal year to liquidate government obligations.

Federal Emergency Management Agency Disaster Relief Fund Obligations and Expenditures for Hurricanes Harvey, Irma, Maria, and California Wildfires through April 30, 2018.

Nevertheless, GAO found that FEMA faced a number of challenges that slowed and complicated its response efforts to Hurricane Maria, particularly in Puerto Rico. Many of these challenges were also highlighted in FEMA's own 2017 hurricane after action report, including:

- the sequential and overlapping timing of the three hurricanes—with Maria being the last of the three—caused staffing shortages and required FEMA to shift staff to the territories that were already deployed to other disasters;

- logistical challenges complicated efforts to deploy federal resources and personnel quickly given the remote distance of both territories; and
- limited preparedness by the U.S. Virgin Islands and Puerto Rico for a Category 5 hurricane and incapacitation of local response functions due to widespread devastation and loss of power and communications led FEMA to assume response functions that territories would usually perform themselves.

The 2017 hurricanes and wildfires highlighted some longstanding issues and revealed other emerging response and recovery challenges. For example, the concurrent timing and scale of the disaster damages nationwide caused shortages in available debris removal contractors and delays in removing disaster debris—a key first step in recovery. In addition, FEMA's available workforce was overwhelmed by the response needs. For example, at the height of FEMA workforce deployments in October 2017, 54 percent of staff were serving in a capacity in which they did not hold the title of "Qualified"—according to FEMA's qualification system standards—a past challenge GAO has identified. FEMA officials noted that staff shortages, and lack of trained personnel with program expertise led to complications in its response efforts, particularly after Hurricane Maria.

Further, federal, state, and local officials faced challenges finding temporary housing for disaster survivors given the extensive damage to available housing in each location. For example, given the widespread damage in Puerto Rico and lack of hotels and other temporary housing, FEMA transported survivors to the mainland United States to stay in hotels. FEMA also used new authorities and procedures to meet the need, such as providing Texas as much as $1 billion to manage its own housing program. However, this approach had not been used or tested in past disasters and state officials noted challenges in managing the program such as staffing shortfalls. State officials further noted challenges in coordinating with FEMA that led to delays in providing assistance to survivors. GAO will continue to monitor these programs.

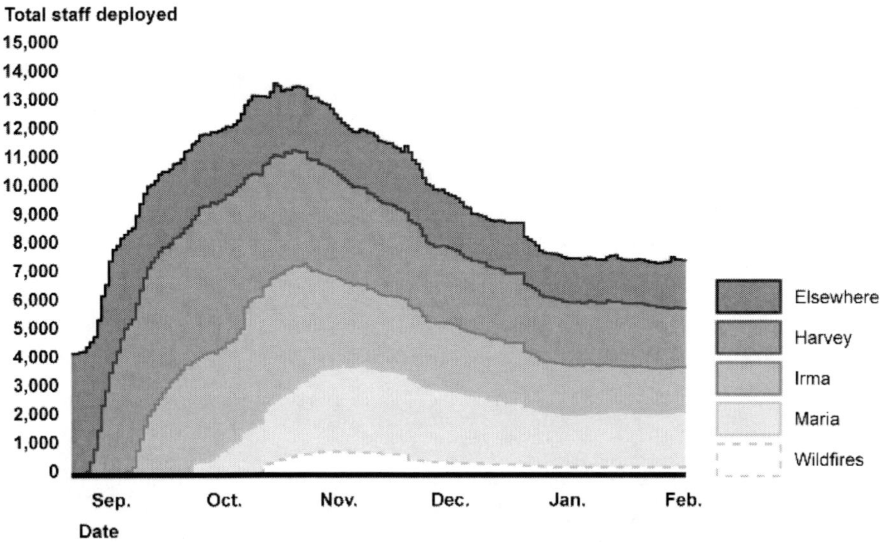

Source: GAO analysis of Department of Homeland Security dat. | GAO-18-472.

Federal Disaster Workforce Deployed at the Height of 2017 Response Activities.

September 4, 2018
Congressional Addressees

In 2017, four near-sequential disasters—Hurricane Harvey, Hurricane Irma, Hurricane Maria, and the California wildfires—created an unprecedented demand for federal disaster response and recovery resources.[1] According to the Federal Emergency Management Agency (FEMA), the 2017 hurricanes and wildfires collectively affected 47 million people—nearly 15 percent of the nation's population—with hurricanes Harvey, Irma, and Maria ranking among the top five costliest hurricanes on record.[2] See Figure 1 for a timeline of these major disasters.[3]

---

[1] The focus of this report is on five geographic areas—Texas, Florida, Puerto Rico, the U.S. Virgin Islands, and California—affected by 4 out of the 137 presidentially disasters declared in 2017.

[2] According to FEMA, the five costliest hurricanes on record are Hurricane Katrina at $161 billion, Hurricane Harvey at $125 billion, Hurricane Maria at $90 billion, Hurricane Sandy at $71 billion, and Hurricane Irma at $50 billion.

[3] A major disaster is any natural catastrophe (including any hurricane, tornado, storm, high water, wind-driven water, tidal wave, tsunami, earthquake, volcanic eruption, landslide, mudslide, snowstorm, or drought), or, regardless of cause, any fire, flood, or explosion, in any part of

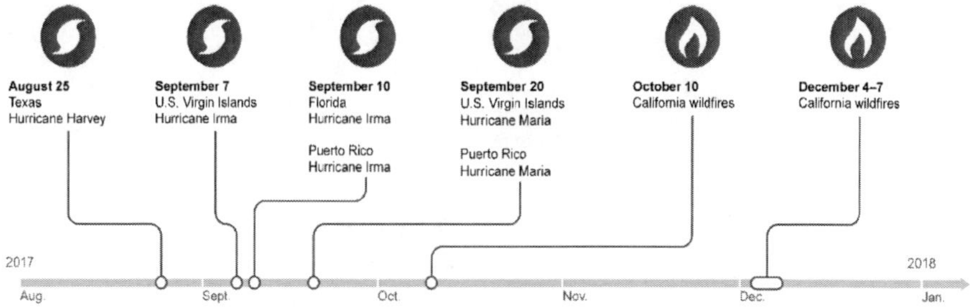

Source: Department of Homeland Security. | GAO-18-472.

Figure 1. Timeline of the 2017 Hurricanes and Wildfires.

We have previously reported that the rising number and costs of disasters and the increasing reliance on the federal government for disaster assistance is a key source of federal fiscal exposure,[4] and that this cost will likely continue to rise as the climate changes.[5] In September 2016, we reported that from fiscal years 2005 through 2014, the federal government obligated over $277 billion for disaster assistance programs and activities. The National Oceanic and Atmospheric Administration estimated that the cumulative damages from weather- and climate-related disasters in 2017 alone cost the United States over $300 billion, making it the costliest year on record. As of June 2018, three supplemental appropriations bills have been enacted, providing over $120 billion in supplemental federal funding for activities related to the 2017 hurricanes and wildfires.[6]

---

the United States, which the president determines causes damage of sufficient severity and magnitude to warrant major disaster assistance to supplement the efforts and available resources of states, local governments, and disaster relief organizations in alleviating damage, loss, hardship, or suffering. See 42 U.S.C. § 5122(2).

[4] GAO, *Fiscal Exposures: Improving Cost Recognition in the Federal Budget*, GAO-14-28 (Washington, D.C.: Oct. 29, 2013).

[5] GAO, *Climate Change: Information on Potential Economic Effects Could Help Guide Federal Efforts to Reduce Fiscal Exposure*, GAO-17-720 (Washington, D.C.: Sept. 28, 2017). Managing fiscal exposure due to climate change has been on our high risk list since 2013, in part, because of concerns about the increasing costs of disaster response and recovery efforts. See GAO, *High-Risk Series: An Update*, GAO-15-290 (Washington, D.C.: Feb. 11, 2015); also http://www.gao.gov/highrisk/limiting_federal_government_fiscal_exposure/why_did_study.

[6] This figure does not include transfers of unobligated balances from prior fiscal years or indefinite appropriations authorized to forgive any outstanding balance owed to the Department of Education under the Historically Black College and University Hurricane

In 2005, Hurricane Katrina became the single largest, most destructive natural disaster in our nation's history causing over 1,800 deaths and an estimated $108 billion in damage. In the wake of Hurricane Katrina, we and others identified several issues of leadership and planning that plagued the response.[7] To address these critiques, Congress passed the Post-Katrina Emergency Management Reform Act of 2006 (Post-Katrina Act).[8] Among other things, the act clarified FEMA's roles and responsibilities as the primary federal agency responsible for disaster preparedness, response, and recovery, and provided additional authorities to federal agencies to address the shortcomings from Katrina.

Since the Post-Katrina Act was enacted in 2006, we have evaluated a range of emergency management issues including federal efforts to implement provisions of the act and improve national emergency preparedness, response, and recovery.[9] We have also evaluated the federal response to other major disasters since that time, most notably the response to and recovery from Hurricane Sandy in 2012. We have made numerous recommendations to FEMA and other federal agencies to strengthen their disaster response efforts, many of which have been implemented or are in the process of being implemented. We discuss some of these recommendations in more detail throughout this chapter.

Given the scale and cost of the 2017 disasters, Congress and others have raised questions about the federal response to these disasters and

---

Supplemental Loan program. Also, the supplemental appropriations provided up to $78.5 million for oversight activities by nine Inspectors General and us related to the expenditure of these funds. Supplemental Appropriations for Disaster Relief Requirements Act, 2017, Pub. L. No. 115-56, div. B, 131 Stat. 1129, 1136 (2017); Additional Supplemental Appropriations for Disaster Relief Requirements Act, 2017 Pub. L. No. 115-72, div. A, 131 Stat. 1224, 1224 (2017); Supplemental Appropriations for Disaster Relief Requirements Act, 2018, Pub. L. No. 115-123, div. B, subdiv. 1, 132 Stat. 64, 65 (2018).

[7] See, for example GAO, *Hurricane Katrina: GAO's Preliminary Observations Regarding Preparedness, Response, and Recovery*, GAO-06-442T (Washington, DC: Mar. 8, 2006).

[8] Pub. L. No. 109-295, tit. VI, 120 Stat. 1355, 1394 (2006); see also 6 U.S.C. §§ 721, 722; 42 U.S.C. § 5144. The provisions of the Post-Katrina Act became effective upon enactment, October 4, 2006, with the exception of certain organizational changes related to FEMA, most of which took effect on March 31, 2007.

[9] Two reports focused specifically on the Post-Katrina Act; see GAO, *Actions Taken to Implement the Post-Katrina Emergency Management Reform Act of 2006*, GAO-09-59R, (Washington, D.C.: Nov. 21, 2008); and GAO, *National Preparedness: Actions Taken by FEMA to Implement Select Provisions of the Post-Katrina Emergency Management Reform Act of 2006,* GAO-14-99R (Washington, D.C.: Nov 26, 2013).

various recovery challenges that have arisen since these disasters. Under the authority of the U.S. Comptroller General to undertake reviews that help inform Congressional oversight, we initiated a review of the federal government's handling of the 2017 hurricanes and wildfires. Specifically, this chapter addresses:

1) our observations of federal and state preparedness and the response coordination for hurricanes Harvey and Irma in Texas and Florida, as well as for the California wildfires;
2) our observations of the federal preparedness and response to hurricanes Irma and Maria in Puerto Rico and the U.S. Virgin Islands; and
3) existing and emerging disaster recovery challenges highlighted by the 2017 hurricanes and wildfires.

In addition, this chapter includes 10 appendices that provide further details and data related to the federal response to the 2017 disasters and various recovery challenges. These appendices cover key issues and challenges that we believe are critical to the federal response and that warrant continued congressional and agency oversight during recovery. This work will include assessments of federal preparedness, planning, response, and recovery efforts. The appendices are:

- Appendix II: Federal Appropriations and FEMA Obligations for the 2017 Hurricanes and California Wildfires
- Appendix III: Federal Response Coordination during the 2017 Hurricanes and California Wildfires
- Appendix IV: Federal Contracting for the 2017 Hurricanes
- Appendix V: FEMA Disaster Workforce Capacity
- Appendix VI: FEMA's Individual Assistance Program
- Appendix VII: Fraud Risk Management in FEMA's Disaster Assistance Programs
- Appendix VIII: Payment Integrity and Prior Identified Requirements for Disaster Relief Funding

- Appendix IX: FEMA's Public Assistance Program
- Appendix X: Disaster Resilience and Hazard Mitigation
- Appendix XI: Department of Defense's Support of Civil Authorities during the 2017 Hurricanes and California Wildfires

To address all three objectives, we analyzed federal laws and FEMA policies, procedures, and guidance specific to emergency management. Specifically, we reviewed select sections of the Post-Katrina Act, including those associated with the establishment of (1) the *National Response Framework*, (2) the Federal Coordinating Officer (FCO) position—the lead federal official in charge of response, (3) Incident Management Assistance Teams (IMAT)—FEMA staff who rapidly deploy to an incident to provide leadership in the identification and provision of federal assistance and federal response capabilities, (4) the surge capacity force; and (5) the Sandy Recovery Improvement Act, particularly those sections associated with FEMA's public assistance program and debris removal responsibilities.[10] Additionally we reviewed the *National Response Framework*, National Disaster Recovery Framework, 2017 National Preparedness Report, and FEMA's 2014-2018 Strategic Plans. We also reviewed relevant information from our prior reports on FEMA's work.[11]

---

[10] The *National Response Framework* is the part of the National Preparedness System established in Presidential Policy Directive 8 that is to be used to manage any type of disaster or emergency response, regardless of scale, scope, and complexity. Specifically, this framework covers actions to save lives, protect property and the environment, stabilize communities, and meet basic human needs following an incident. Response also includes the execution of emergency plans and actions to support short-term recovery. Department of Homeland Security, Federal Emergency Management Agency, *National Response Framework, Third Edition* (Washington, D.C.: June 2016). The surge capacity force is a cadre of non-FEMA federal employees who augment FEMA's disaster response and recovery efforts.

[11] GAO-14-28; GAO-17-720; GAO-15-290; GAO-06-442T; GAO-09-59R; GAO-14-99R; *2017 Disaster Contracting: Observations on Federal Contracting for Response and Recovery Efforts*. GAO-18-335. (Washington, D.C.: Feb. 28, 2018); GAO, *Disaster Recovery: FEMA's Public Assistance Grant Program Experienced Challenges with Gulf Coast Rebuilding*, GAO-09-129 (Washington, D.C.: Dec. 18, 2008); GAO, *Hurricane Sandy: An Investment Strategy Could Help the Federal Government Enhance National Resilience for Future Disasters*, GAO-15-515 (Washington, D.C.: July 30, 2015); GAO, *Federal Emergency management Agency: Workforce Planning and Training Count be Enhanced by Incorporating Strategic Management Principles*, GAO-12-487 (Washington, D.C.: Apr. 26, 2012); GAO, *Federal Emergency Management Agency: Additional Planning and Data*

Further, we analyzed key data from FEMA's financial management, workforce, and emergency operations systems for the period August 2017 through January 2018—the highest period of disaster response activity for the federal government—and obtained updates from FEMA through June 2018. We interviewed officials at FEMA headquarters about their data quality control procedures, reviewed existing information about data systems—particularly data definitions and data validation, conducted electronic testing and reviewed the data for obvious errors and omissions to ensure that all data were sufficiently reliable for the purposes of our reporting objectives, as described in appendix I.[12] See Related GAO Products for a full list of our products related to each appendix contained in this chapter.

Moreover, in October and November 2017, we conducted site visits to hurricane damaged areas in the greater Houston area, throughout southern Florida, in San Juan, Puerto Rico, and St. Croix, U.S. Virgin Islands, and visited FEMA's joint field offices (JFO) —multiagency coordination centers established near disaster sites for coordinating major disaster response and recovery efforts—for Hurricane Harvey, located in Austin Texas; Hurricane Irma, located in Orlando Florida; and hurricanes Irma and Maria, located in San Juan and St. Croix. At these locations, we interviewed FEMA's on-site leadership and conducted site visits to FEMA Disaster Response Centers. Further, we conducted interviews with FEMA's on-site leadership responsible for the response and recovery efforts in California. We also interviewed emergency management officials or their designees in each disaster-affected state and territory as well as local government officials from eight municipalities in Texas, Florida, and California to gain their insights and perspectives on the federal response to the hurricanes and wildfires in their respective states and territories. We selected the cities and counties whose officials we interviewed based on their geographic proximity to the disaster-affected sites we were already

Collection Could Help Improve Workforce Management Efforts, GAO-15-437 (Washington, D.C.: July, 8, 2015).

12 FEMA has 10 regional offices located across the continental United States See appendix XII for the location of each as well as the states each regional office is responsible for collaborating with to administer FEMA programs.

visiting, and their availability. The findings from these interviews cannot be generalized to all disaster-affected states, however, they provided valuable insights about their respective state's and the federal response to and recovery from the disasters. We also conducted interviews with Department of Defense officials who assisted FEMA in its response efforts. Further information on our scope and methodology can be found in appendix I.

We are not making recommendations in this chapter, but it is part of a body of work on related issues across federal departments as those discussed in appendix II through XI of this chapter, and we will further assess these issues moving forward, making recommendations, as appropriate, once this work is completed.

We conducted this performance audit from September 2017 to September 2018 in accordance with generally accepted government auditing standards. Those standards require that we plan and perform the audit to obtain sufficient, appropriate evidence to provide a reasonable basis for our findings and conclusions based on our audit objectives. We believe that the evidence obtained provides a reasonable basis for our findings and conclusions based on our audit objectives.

BACKGROUND

Disaster Response Roles and Responsibilities

Disaster response can involve many federal, state, territorial, tribal, private sector, and nongovernmental entities. The *National Response Framework* describes how the federal government, states and localities, and other public and private sector institutions should respond to disasters and emergencies. For example, state, local, tribal and territorial governments are to play the lead roles in disaster response and recovery. Local emergency agencies—police, firefighters, and medical teams—are to be the first responders in a disaster or emergency.

Emergency Support Functions

Emergency support functions (ESF) are the federal government's primary coordinating structure for building, sustaining, and delivering response capabilities. There are 14 ESFs, organized by specific functional areas for the most frequently needed capabilities during an emergency. ESFs are designed to coordinate the provision of related assets and services by federal departments and agencies. For each of the 14 ESFs, a federal department or agency serves as the designated ESF coordinator:

ESF #1: Transportation (Department of Transportation)

ESF #2: Communications (Department of Homeland Security National Communication System)

ESF #3:Public Works and Engineering (Department of Defense/U.S. Army Corps of Engineers)

ESF #4:Firefighting (US Forest Service) ESF #5: Information and Planning (Federal Emergency Management Agency) ESF #6: Mass Care, Emergency Assistance, Housing, and Human Services (Federal Emergency Management Agency) ESF #7: Logistics Management (General Services Administration and Federal Emergency Management Agency)

ESF #8: Public Health and Medical Services (Department of Health and Human Services)

ESF #9: Search and Rescue (Federal Emergency Management Agency) ESF #10: Oil and Hazardous Materials Response Environmental Protection Agency)

ESF #11: - Agriculture and Natural Resources (Department of Agriculture)

ESF #12: Energy (Department of Energy)

ESF #13: Public Safety and Security (Department of Justice)

ESF# 14—No longer used

ESF #15: External Affairs (Department of Homeland Security)

Source: GAO, *Emergency Preparedness: Opportunities Exist to Strengthen Interagency Assessments and Accountability for Closing Capability Gaps*, GAO-15-20 (Washington, D.C.: Dec. 4, 2014).I GAO-18-472.

Federal agencies can become involved in responding to a disaster when effective response and recovery are beyond the capabilities of the state and affected local governments. In such cases, the Robert T. Stafford Disaster Relief and Emergency Assistance Act (Stafford Act), permits the President to declare a major disaster in response to a request by the governor of a state or territory or by the chief executive of a tribal government.[13] Such a declaration is the mechanism by which the federal government gets involved in funding and coordinating response and recovery activities.[14] Under the *National Response Framework*, the Department of Homeland Security (DHS) is the federal department with primary responsibility for coordinating disaster response, and within DHS, FEMA has lead responsibility. The Administrator of FEMA serves as the principal adviser to the President and the Secretary of Homeland Security regarding emergency management.[15]

Once a major disaster is declared, states, territories, and tribes may obtain federal assistance through the Disaster Relief Fund (DRF).[16] In general, response and recovery activities that FEMA coordinates under the Stafford Act are funded from the DRF. See appendix II for more information on DRF spending in response to the 2017 disasters.

In addition to DHS, at least 29 other federal agencies carry out disaster assistance programs and activities. The *National Response Framework* identifies 14 emergency support functions (ESFs)—such as communication, transportation, and energy—and designates a federal department or agency as the coordinating agency for each function. For example, provision of assets and services related to public works and

---

[13] 42 U.S.C. § 5170.

[14] *Presidential Policy Directive-8 National Preparedness* (PPD-8) establishes a national preparedness system made of an integrated set of guidance, programs, and processes designed to strengthen the security and resilience of the United States through systematic preparation for the natural and human-caused threats that pose the greatest risk. This system breaks preparedness activities into five different lines of effort—prevention, protection, mitigation, response, and recovery—each of which requires a separate planning framework.

[15] 6 U.S.C. § 313(c)(4).

[16] The DRF is the primary source of federal disaster assistance for state and local governments when a disaster is declared. The DRF is appropriated no-year funding, which allows FEMA to fund, direct, coordinate, and manage response and recovery efforts—including certain efforts by other federal agencies and state and local governments, among others—associated with domestic disasters and emergencies.

engineering, such as temporary roofing or power, are coordinated by the U.S. Army Corps of Engineers (USACE), a component of the Department of Defense (DOD). See appendix III for more information on the 14 ESFs and their assigned coordinating agencies.

Finally, the federal government also works with private-sector businesses and nongovernmental organizations such as the Red Cross, Salvation Army, and other voluntary organizations to provide food, shelter, and essential needs to survivors.

FEMA's Disaster Response Mechanisms

FEMA has multiple mechanisms by which to help coordinate and deliver the federal government's response to disasters. Among those are:

- direct provision of assistance. When a state, tribe, or territory that has received a major disaster declaration requests federal assistance, FEMA can provide that assistance directly in various forms, such as meals, water, or tarps.
- mission assignment to other agencies. FEMA coordinates disaster response efforts through mission assignments—work orders it issues that direct another federal agency to utilize its authorities and the resources granted to it under federal law in support of direct assistance to state, local, tribal, and territorial governments.[17] For example, FEMA often requests medical teams from the Department of Health and Human Services and logistical support from DOD.[18]

---

[17] 42 U.S.C. § 5192(a)(1). The Stafford Act authorizes the President to direct any federal agency, with or without reimbursement, to utilize its authorities and the resources granted to it under federal law in support of state and local response efforts for emergencies. This tasking authority, delegated to the FEMA Administrator, is carried out through a mission assignment.

[18] While DOD's primary mission is to defend the nation, the department is often asked to play a prominent role in supporting civil authorities and must be prepared to provide rapid response when called upon during disasters and declared emergencies (both natural and

- distribution of donations. FEMA can accept and distribute donations and gifts of services, money, or property to alleviate the suffering and damage caused by disasters.[19]
- interagency agreements. FEMA can also acquire supplies or services from other government agencies by executing an interagency agreement with those agencies.
- procurement of supplies and services from contractors. FEMA and other federal agencies support disaster response and recovery by procuring goods and services through contracts.

To provide disaster relief and recovery assistance, federal departments may have to solicit, award, and administer contracts. As of March 31, 2018, federal departments had obligated approximately $7.1 billion for contracts in support of hurricanes Harvey, Irma, and Maria.[20] The Post-Katrina Act, which addressed various shortcomings identified in the preparation for and response to Hurricane Katrina, included provisions to update FEMA's contracting practices.[21]

---

human-caused). DOD provides such support through its Defense Support of Civil Authorities mission.

[19] See 42 U.S.C. § 5201(b). According to FEMA, acceptance of gifts is subject to ethical and operational constraints on a case-by-case basis. For example, FEMA Directive 112- 13 establishes the process for accepting gifts from domestic sources and requires authorized agency officials to determine whether or not the gift reflects poorly on the agency, compromises the agency's integrity, attaches prohibited conditions on the gift or requires the agency to act outside its mission and duties, requires the expenditure of appropriated funds, provides the donor with some benefit, or creates a conflict of interest or the appearance of a conflict of interest.

[20] For the purposes of this report and appendix IV, contract obligations include obligations against what the Federal Procurement Data System-Next Generation (FPDS-NG) categorizes as definitive vehicles (definitive contracts and purchase orders that have a defined scope of work that do not allow for individual orders under them), and against what FPDS-NG categorizes as indefinite delivery vehicles (orders under the Federal Supply Schedule, orders/calls under blanket purchase agreements, orders under basic ordering agreements, orders under government-wide acquisition contracts, and orders under other indefinite delivery vehicles, such as indefinite delivery, indefinite quantity contracts).

[21] Pub. L. No. 109-295, §§ 601-699, 120 Stat. at 1394-1463. We have previously reported in 2015 that FEMA had not fully implemented the Post-Katrina Act's statutorily required contracting reforms following Hurricane Katrina; see GAO-15-783. We made eight recommendations to the FEMA Administrator and one recommendation to the Secretary of Homeland Security, three of which remain open.

One of these provisions requires that FEMA identify and establish contracts prior to a disaster for goods and services that are typically needed during a disaster response—known as "advance" or "pre-positioned" contracts.[22] We are currently conducting more detailed reviews of federal contracting related to the 2017 disasters, including the wildfires. For more information on federal disaster contracting for the 2017 hurricanes, see appendix IV. Figure 2 shows the various mechanisms by which the federal government provides disaster response support.

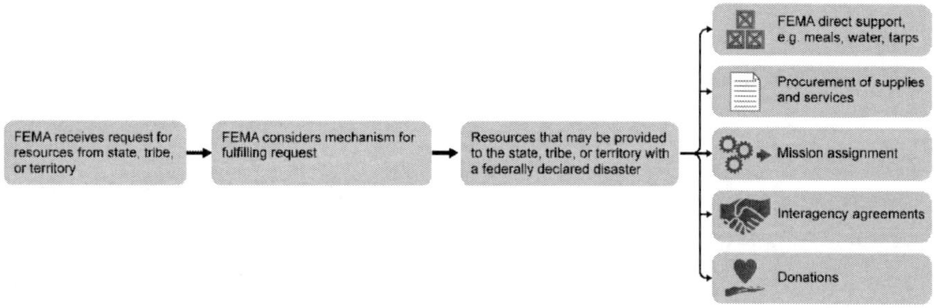

Source: Federal Emergency Management Agency. | GAO-18-472.

Figure 2. Mechanisms Used by Federal Emergency Management Agency (FEMA) to Coordinate and Deliver Disaster Response.

Federal Disaster Workforce

The federal disaster workforce is designed to scale up or down depending on the timing and magnitude of disasters, and includes the following categories of employees:

- Title 5 employees.[23] These permanent and temporary employees make up FEMA's day-to-day workforce and are responsible for administering the agency's ongoing program activities. During disasters, these employees can be deployed as needed. Examples of

[22] In February 2018, we issued a report on our initial observations of federal contracting for response and recovery from the 2017 hurricanes. See GAO-18-335.

[23] Generally, Title 5 refers to the section of the United States Code that establishes the law for managing human resources in the federal government.

Title 5 employees include logistics specialists, contract officers, and budget analysts.

Incident Management Assistance Teams

The primary mission of an incident management assistance team (IMAT) is to rapidly deploy to a disaster area, provide leadership in identifying and providing federal assistance, and coordinate and integrate inter-jurisdictional response to support an affected state, tribe or territory.

IMATs are made up of Federal Emergency Management Agency (FEMA) staff in areas such as operations, logistics, planning, and finance and administration. FEMA has two types of IMAT teams—three national teams comprised of 32 personnel and 13 regional teams comprised of 12 personnel.

Source: GAO, *Disaster Response: FEMA Has Made Progress Implementing Key Programs, but Opportunities for Improvement Exist*, GAO-16-87 (Washington D.C. Feb 5, 2016). I GAO-18-472.

- Stafford Act employees.[24] Stafford Act employees provide support for disaster-related activities and augment FEMA's disaster workforce at facilities, regional offices, and headquarters. Stafford Act employees include a Cadre of On Call Response/Recovery Employees who are temporary employees with 2- to 4-year appointments and can be deployed to fulfill any role specifically related to the incident for which they are hired and qualified. IMAT staff are Cadre of On Call Response/Recovery Employees. They also include reservists, who work on an intermittent basis and are deployed as needed to fulfill incident management roles within their cadre function.[25]

[24] See 42 U.S.C. § 5149(b)(1).

[25] Reservists' activities can include interviewing disaster survivors; conducting and verifying damage assessments; providing administrative, financial, and logistical support; and performing a wide variety of other tasks as identified by staffing needs and operational requirements.

- Surge Capacity Force. The Surge Capacity Force supplements FEMA's disaster workforce in a major disaster and consists of volunteers who are employees of DHS components, such as the Transportation Security Administration and U.S. Secret Service, as well as employees of other federal agencies, as authorized by the Post-Katrina Act.[26] Surge Capacity Force volunteers are deployed to disaster sites for a maximum of 3 months.
- FEMA Corps. FEMA Corps is a team-based national service program operated by AmeriCorps in partnership with FEMA. Members are not FEMA employees, but are deployed to augment FEMA's workforce for disaster readiness, preparedness, response, and recovery work under the supervision of FEMA staff. FEMA staff are responsible for developing projects for FEMA Corps members and providing technical supervision at project sites. FEMA Corps members are generally 18 to 24 years old and serve 10-month terms.

In addition to these four types of employees, FEMA hires locally and employs other personnel, such as contractors, to provide a variety of forms of assistance and services to meet disaster preparedness, response, and recovery needs, such as debris removal. Prior to Hurricane Harvey in August 2017, the federal disaster workforce, including Surge Capacity Force and FEMA Corps, was 24,040. As of January 2018, the federal disaster workforce had grown to 33,041, as shown in Figure 3.[27] For more information on FEMA's disaster workforce, see appendix V.

---

[26] 6 U.S.C. § 711(b).

[27] FEMA's disaster workforce, which is a component of the overall federal disaster workforce, was 11,213 prior to Hurricane Harvey, and had grown to 11,980 as of January 2018. According to FEMA officials, not all FEMA personnel can be deployed to a disaster. Those who cannot be deployed provide support to FEMA headquarters or regional offices, and the National Response Coordination Center. These personnel are part of FEMA's force strength, but not part of the disaster workforce. Additionally, several factors can affect the availability of the federal disaster workforce, including whether employees are on leave. Also, FEMA does not have direct oversight regarding the availability of the Surge Capacity Force, contractors, other federal agencies, or FEMA Corps as they are not FEMA employees.

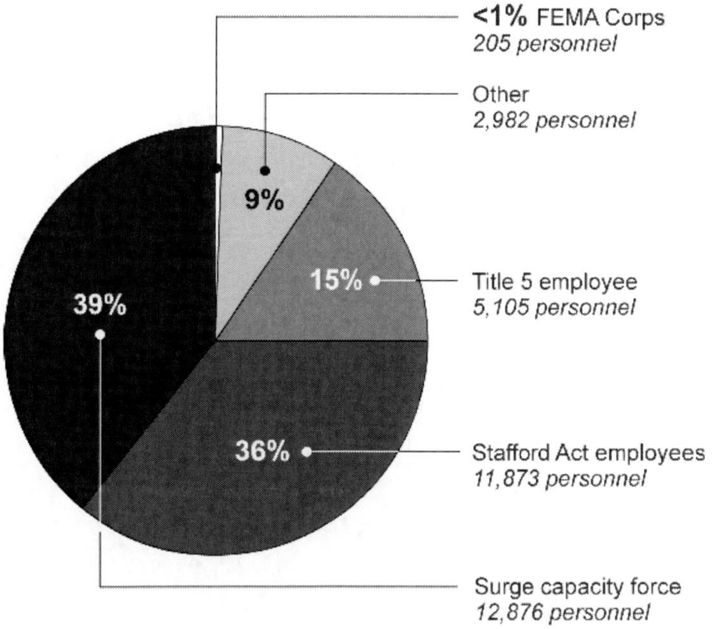

Source: GAO analysis of Department of Homeland Security data. | GAO-18-472.

Note: “Other” includes the following four workforce categories: permanent part-time, temporary part-time, temporary incident, and political appointees. Federal Emergency Management Agency (FEMA) Corps is a team-based national service program operated by AmeriCorps National Civilian Community Corps in partnership with FEMA. Members are not FEMA employees, but are deployed to augment FEMA’s workforce for disaster readiness, preparedness, response, and recovery work under the supervision of FEMA staff. Title 5 employees are permanent and temporary employees who make up FEMA’s day-to-day workforce and are responsible for administering the agency’s ongoing program activities. Stafford Act employees include Cadre of On Call Response/Recovery Employees who are temporary employees with 2- to 4-year appointments who can be deployed to fulfill any role specifically related to the incident for which they are hired and qualified. The Surge Capacity Force supplements FEMA’s disaster workforce in a major disaster and consists of volunteers who are employees of the Department of Homeland Security components and other federal agencies.

Figure 3. Federal Disaster Workforce as of January 31, 2018.

Overview of FEMA and Federal Disaster Recovery Programs

After a disaster strikes, the response phase typically lasts for days or weeks, depending on the impact and complexity of the disaster and eventually transitions into recovery operations. As with response, a number of federal departments and agencies may assist with various forms of disaster recovery assistance to individuals and state, local, tribal, and territorial governments. While this chapter focuses primarily on recovery programs that FEMA delivers, we are conducting work on other federal programs and issues, including use of the Community Development Block Grant and the National Flood Insurance Program for recovery from the disasters, as part of our collective body of work on the 2017 disasters.

FEMA provides three principal forms of disaster recovery funding assistance—Individual Assistance (IA), Public Assistance (PA) and Hazard Mitigation.[28]

Individual Assistance

FEMA's IA program provides financial assistance directly to survivors for the necessary expenses and serious needs that cannot be met through insurance or low-interest loans, such as temporary housing assistance, counseling, unemployment compensation, or medical expenses (for more information on FEMA's IA Program see appendix VI). Part of its mission is to provide this assistance quickly. In response to our previously identified weaknesses, FEMA has taken steps to improve its ability to do so while protecting government resources. For example, in March 2018, FEMA reported that the agency had collaborated with the Social Security Administration to assess the feasibility of a direct data exchange with the Administration for the purpose of identifying recipients using Social Security numbers that were ineligible or likely belonged to deceased

[28] In addition to the three forms of funding assistance, the DRF also provides funding for (1) Fire Management Assistance grants to state, local and tribal governments for the mitigation, management, and control of fires; (2) Mission Assignment which allows FEMA to issue task orders directing other federal agencies to provide direct assistance to disaster affected states, tribes, and territories; and (3) Administration to cover FEMA's costs for supporting the delivery of disaster assistance.

individuals. It also reported that it had taken steps to more reliably determine eligibility for its Individuals and Households Program based on compliance with flood-insurance requirements. The agency took these actions in an effort to address recommendations we made in 2015 to help prevent improper payments.[29] See appendix VII for information on FEMA's actions to manage fraud risk related to its disaster assistance programs. Further, given the significant costs of these four disasters to the federal government, it is important that federal agencies tasked with response and recovery programs pay particular attention to internal controls and payment integrity issues related to disaster relief. See appendix VIII for information on issues we identified in our prior work related to disaster relief payment integrity.[30]

Public Assistance

FEMA's PA program provides supplemental federal disaster grant assistance to state, local, tribal, and territorial governments, and certain types of private nonprofit organizations for debris removal, emergency protective measures, and the restoration of disaster-damaged, publicly-owned facilities and the facilities of certain private nonprofit organizations. The PA program also encourages protection of these damaged facilities from future events by providing assistance for hazard mitigation measures. The program—which represents the largest share of federal aid from the

---

[29] GAO-15-15.

[30] Effective fraud risk management can help ensure that federal disaster assistance programs serve their intended purpose, taxpayer dollars are spent effectively, and government assets are safeguarded. Since 2014, when we last reported on FEMA's implementation of controls to help prevent potentially improper or fraudulent payments in the IA program, we issued *A Framework for Managing Fraud Risks in Federal Programs* (Fraud Risk Framework). The Fraud Risk Framework provides a comprehensive set of leading practices that serve as a guide for agency managers to use when developing efforts to combat fraud in a strategic, risk-based way. In addition, the Fraud Reduction and Data Analytics Act of 2015, enacted in June 2016, requires the Office of Management and Budget to establish guidelines for agencies for implementing control activities to prevent, detect, and respond to fraud, including improper payments, and to incorporate the Fraud Risk Framework's leading practices into the guidelines. See GAO, *A Framework for Managing Fraud Risks in Federal Programs*, GAO-15-593SP (Washington, D.C.: July 28, 2015). Additionally, in April, 2018 we started a review focused on select agencies' design and implementation of key internal control activities related to preventing and detecting improper payments of disaster relief and recovery funding in response to the 2017 hurricanes and wildfires.

Disaster Relief Fund—is administered through a partnership between FEMA and the state, tribal or territorial grantee, which provides funding to local or tribal entities who are the subrecipients of a PA grant award. Thus, it entails an extensive paperwork and review process between FEMA and grantee officials based on specific eligibility rules that outline the types of damage that can be reimbursed by the federal government and steps that federal, state, and local governments must take in order to document eligibility. We have identified a number of past challenges affecting various aspects of the PA program. To address these various challenges, we made a number of recommendations, and FEMA has taken or is taking various actions to address them. For example, as of January 2018, FEMA officials had begun incorporating experiences and lessons learned from the 2017 hurricane season and planned to reevaluate the appropriate number of staff needed in the PA workforce, and present recommendations to senior leadership. Officials also reported completing activities to develop disaster-specific mitigation performance measures that align with strategic goals, and analyzed available data to identify the drivers of mitigation in events of various sizes. The agency took these actions in response to recommendations we made in 2018 to complete a workforce staffing assessment that identifies the appropriate number of staff to implement a new PA delivery model nationwide, and to develop performance measures for the new delivery model that better align with the agency's strategic goal for hazard mitigation.[31]

The Sandy Recovery Improvement Act of 2013 (SRIA) authorized the use of alternative procedures in administering the PA program, thereby providing new flexibilities to FEMA, states, and local governments for debris removal, infrastructure repair, and rebuilding projects using funds from this program.[32] The stated goals of the alternative procedures are to reduce the costs to the federal government, increase flexibility in the

[31] GAO-18-30.

[32] SRIA amended the Stafford Act by adding Section 428, which authorized FEMA to approve Public Assistance program projects under the alternative procedures provided by that section for any presidentially-declared major disaster or emergency. This section further authorized FEMA to carry out the alternative procedures as a pilot program until FEMA promulgates regulations to implement this section. Pub. L. No. 113-2, div. B, § 1102(2), 127 Stat. 39, amending Pub. L. No. 93-288, tit. IV, § 428 (codified at 42 U.S.C. § 5189f).

administration of the PA program, expedite the provision of assistance under the program, and provide financial incentives for recipients of the program for the timely and cost-effective completion of projects.[33] Alternative procedures for permanent work are designed to give jurisdictions more flexibility in determining how, where, and what to rebuild, particularly after incurring significant damage. Applicants may choose to combine multiple critical facilities and rebuild them in a manner that makes them less likely to incur future disaster damages. In 2013 FEMA began a pilot program to utilize the alternative procedures for debris removal and permanent work projects in the recovery from Hurricane Sandy in New York and New Jersey, as of April 2018, FEMA reported that 29 percent of New York's permanent work projects are under the alternate procedures—approximately $8.6 billion. However, no state or territory has used alternative procedures for 100 percent of their permanent work projects. According to Puerto Rico's Draft Recovery Plan issued in July 2018, the Commonwealth estimated costs for permanent work projects ranges from $26.7 billion to $37.4 billion.[34]

In 2015, FEMA awarded a contract for program support to help implement a redesigned PA program. FEMA officials told us that the redesigning effort was primarily focused on specializing roles, segmenting the work, standardizing processes, and consolidating resources. It also included developing a new information system (PA Grants Manager and Grants Portal) to better maintain and share grant documentation. Taken together, according to officials, these efforts represent FEMA's "new delivery model" for the PA program, and, represents a significant process and cultural shift towards a streamlined and standardized way of delivering PA.[35]

---

[33] See 42 U.S.C. § 5189f(c).

[34] DHS, Preliminary Draft: *"Transformation and Innovation in the Wake of Devastation: An Economic and Disaster Recovery Plan for Puerto Rico."* July 9, 2018.

[35] According to FEMA officials, the new delivery model is a re-engineering of the previous process, given that the laws, regulations, and policies underlying the PA program were not changed. In addition to creating a new online information system—Grants Portal—for applicants to develop and submit their grant applications and associated documents, key changes brought about by the re-engineering effort included (1)identifying opportunities for hazard mitigation earlier in the process; creating consolidated resource centers to

Applicable Authorities Granted by the Post Katrina Emergency Management Reform Act of 2006 to Enhance Disaster Response:

Improved ability to provide support to states and tribes ahead of a disaster: The Federal Emergency Management Agency (FEMA) gained statutory authority to provide accelerated federal assistance and federal support where necessary to save lives, prevent human suffering, or mitigate severe damage, which may be provided in the absence of a specific request.

Development of a National Disaster Recovery Framework (NDRF): The NDRF defines coordination structures, leadership roles and responsibilities, and guidance for federal agencies, state, local, tribal, and territorial, governments, and other partners involved in disaster planning and recovery.

Establishment of Incident Management Assistance Teams: Rapid response teams are able to deploy within 2 hours and arrive at an incident within 12 hours to support the local incident commander. The teams support the initial establishment of a unified command and provide situational awareness for federal and state decision makers crucial to determining the level and type of immediate federal support that may be required.

Enhanced partnerships with the private sector: As part of this effort, FEMA established the National Business Emergency Operations Center that serves as a clearinghouse for two-way information sharing between public and private sector stakeholders in preparing for, responding to, recovering from, and mitigating disasters.

Source: FEMA. I GAO 18 472.

In September 2017, FEMA decided to begin using the new delivery model nationwide for all subsequent declared disasters, including hurricanes Harvey and Irma in Texas and Florida and the wildfires in

---

standardize and centralize PA staff responsible for managing grant applications; and creating new specialized positions, such as hazard mitigation liaisons, program delivery managers, and site inspectors, to ensure more consistent guidance to applicants. Other changes to the process included enhancing outreach to applicants during the "exploratory call"—the first contact between FEMA and local officials—and during the first in-person meeting, called the "recovery scoping meeting." FEMA also revised decision points during the process when program officials can request more information from applicants, and applicants can review and approve the completion of project development steps.

California. However, for hurricanes Irma and Maria in Puerto Rico and the U.S. Virgin Islands, FEMA is utilizing the PA alternative procedures model. According to FEMA officials, Puerto Rico was already in the process of implementing this model in response to prior disaster events. See appendix IX for more information on FEMA's PA Program.

Selected Examples of Response Resources and Support Provided to Texas, Florida and California

Total Number of Mission Assignments as of January 31, 2018

Hurricane Harvey: 164 Hurricane Irma: 212

Texas: In response to the historic flooding caused by Hurricane Harvey in Texas, multiple agencies—including U.S. Customs and Border Protection, U.S. Coast Guard, and the U.S. Forest Service, Department of Fish and Wildlife—executed mission assignments to provide boating equipment to move up to 20,000 survivors.

Florida: In response to Hurricane Irma in Florida, Housing and Urban Development executed a mission assignment to support multiple programs providing shelter to disaster survivors. For example, assessing and coordinating assistance to elderly populations in the Transitional Shelter Assistance Program, coordinating with Public Housing Authorities on timelines for repairs to damaged units; and efforts to ensure Fair Housing Act compliance.

California: In response to the California wildfires, DOD executed mission assignments to provide commodity support (e.g., 5 million AA batteries) and access to military bases to stage ambulances and their crews, as well as housing support for federal personnel in various agencies.

Source: GAO analysis of FEMA's Web-based Emergency Operations Command data and Defense Logistics Agency data. I GAO 18 472.

Hazard Mitigation Grant Program

This program is designed to improve community resilience—the ability to prepare and plan for, absorb, recover from, and more successfully adapt to disasters—to future disasters during recovery. The program funds a wide range of projects, such as purchasing properties in flood-prone areas, adding shutters to windows to prevent future damage from hurricane winds and rains, and rebuilding culverts in drainage ditches to prevent future flooding damage. In light of our identification of the rise in the number—and the increase in severity— of disasters as a key source of federal fiscal exposure, we and others have advocated hazard mitigation and resiliency to help limit the nation's fiscal exposure.[36] In 2015, we identified challenges in effectively incorporating mitigation into PA projects and grant guidance during the recovery from Hurricane Sandy in the northeastern United States.[37] We recommended that FEMA assess the challenges and implement corrective actions as needed, and that the Mitigation Framework Leadership Group—created to help coordinate hazard mitigation efforts of relevant federal, state, local, territorial, and tribal organizations—establish an investment strategy to identify, prioritize, and implement federal investments in disaster resilience. FEMA concurred with our recommendations and is taking steps to implement them. In appendix X we identify several challenges the communities impacted by the 2017 disasters face and opportunities for resiliency. We have plans to examine these challenges and opportunities more broadly in an upcoming review.

Federal and State Coordination Efforts Facilitated Preparedness and Response to the 2017 Disasters in Texas, Florida, and California

FEMA coordinated closely with Texas, Florida, and California emergency management officials and other federal, local, and volunteer

---

[36] GAO-15-515. Hazard mitigation is any sustained action taken to reduce or eliminate long-term risk to people and property from natural hazards and their effects. In addition to the Hazard Mitigation Grant Program, FEMA may also fund hazard mitigation projects related to the damaged facilities receiving PA funding pursuant to section 406 of the Stafford Act, as amended. 42 U.S.C. § 5172; 44 C.F.R. § 206.226.

[37] GAO-15-515.

emergency partners to implement various emergency preparedness actions prior to the 2017 disasters, in each state, and to respond to these disasters. According to FEMA and state officials, these actions helped overcome a number of challenges they faced such as deploying a sufficient and adequately-trained disaster workforce and removing debris in a timely manner after the hurricanes and wildfires. These efforts also show progress made since the 2006 Post-Katrina Act, which reflected various themes we identified when reviewing the 2005 federal response to Hurricane Katrina, such as the importance of clear procedures for national response activities, advance planning, and robust training and exercise programs.[38]

Preparedness Activities in Texas, Florida, and California Helped Strengthen the Response

FEMA and state officials in Texas, Florida, and California took certain key actions in advance of the 2017 disasters that enabled them to more effectively provide assistance following each disaster. Specifically, according to FEMA's FCO for Hurricane Harvey, Texas had learned from its experience with prior disasters to ensure that personnel and resources were in place before the hurricane arrived. In accordance with the Post-Katrina Act, Texas requested that FEMA deploy IMAT staff prior to hurricane landfall. Subsequently, the IMAT set up centers to distribute meals and water and prepared federal agencies so that response teams and resources were ready to go upon landfall. Before landfall, there were four regional IMAT teams and a national IMAT team already in place. FEMA also dispensed mission assignments to DOD and the U.S. Coast Guard, among other agencies, with ESF responsibilities to carry out missions as needed.

Similarly, Florida's governor requested, and the President approved, a pre-landfall emergency declaration on September 5, 2017, for all 67 counties in Florida. The pre-declaration authorized FEMA to set up

[38] See Pub. L. No. 109-295, tit. VI, 120 Stat. 1355, 1394 (2006).

emergency berms, pre-position supplies, and take other key preparation steps such as sheltering and evacuation support, according to Florida emergency management officials. Florida, like Texas, also requested the pre-positioning of FEMA IMAT response teams. According to FEMA officials, Florida set up an incident operations center in Tallahassee and response personnel were located there before the hurricane made landfall.

In addition, prior to landfall, FEMA set up state-driven task forces in Texas for sheltering, family reunification, and feeding, among other things, to provide support at the local level. For example, the feeding task force coordinated with Feeding America and other nonprofits that provided food and nutrition while the reunification task force worked with the National Center for Missing and Exploited Children to reunite family members who were separated.

According to IMAT officials, they conducted regular emergency response exercises with the states leading up to the hurricanes which better trained and prepared them to coordinate during actual response efforts.

In California, FEMA and state Office of Emergency Services officials credited the ability to quickly and effectively coordinate with federal partners and respond to the wildfires to the state's past emergency preparedness experience and capacity.

Coordination Systems and Activities Helped Build Relationships That Facilitated Response in Texas, Florida, and California

FEMA and state officials in Texas, Florida, and California all described response coordination systems and activities that helped build relationships among federal, state, and local partners that are crucial to an effective response. The emergency management community has long recognized the importance of building relationships before a disaster. As a former FEMA Administrator stated, the worst time to exchange business cards is during a disaster. For example, in Texas and California, those relationships were primarily formed from regular meetings and the exercises that FEMA conducts with state emergency response partners.

Specifically, FEMA officials at the Joint Field Office in Austin, Texas told us that all emergency response entities in Texas work together via an interagency coordination group, which they credited for enhancing the relationship between these entities.

Texas, Florida and California officials also described that having state agency staff embedded in FEMA's organization, and vice versa, was helpful. In each state, FEMA employees were embedded in state offices prior to the disasters, and in doing so had developed a close working relationship with state emergency management personnel. In Texas, embedded state emergency staff provided training, emergency exercises, and advice to local jurisdictions. Texas Division of Emergency Management staff told us that they are in daily contact with local officials year-round, which enhances coordination in times of disaster. According to the California Office of Emergency Services (OES) Director, the pre-existing relationship between California OES, the California governor's office, and FEMA's Region IX office—the FEMA regional office with oversight for the state of California—allowed California to approach the response to the wildfires as a team, with clear roles and responsibilities among the state, the FEMA region and other partners like the National Guard. According to FEMA officials, Region IX staff were members embedded with the state of California for almost 60 days following the start of the wildfires. By colocating from the start, FEMA and California's OES were able to work collaboratively in decision making and setting priorities, according to the California OES director. The California OES director agreed with FEMA officials at the Joint Field Office in Austin who said that co-location facilitated decision-making by reducing the bureaucracy that would typically be involved in reaching out to partners spread out in different regions of the state and country.[39]

In addition to forming a close working relationship with FEMA, officials from all three states described close preexisting working relationships with volunteer organizations that play a significant role in

---

[39] In April 2018, we initiated a separate review of the California wildfires that will more fully examine the federal response to the wildfires, including the role played by FEMA and other relevant federal agencies

disaster response in their states. For example, a FEMA official with responsibility for coordinating volunteer partnerships stated that in Texas, volunteer organizations are treated as full and equal partners, which facilitated volunteer partners' contribution to the response and recovery. Similarly, officials from California and FEMA officials in Florida told us that nonprofits have made significant contributions to response and recovery. Figure 4 shows an example of such contributions following Hurricane Irma.

Although state and FEMA officials largely described a well-coordinated and successful response to the hurricanes and wildfires, all three states experienced a number of challenges—most notably with deployment of a sufficient and adequately-trained FEMA disaster workforce and delays with debris removal—which we describe later in this chapter.[40]

Source: GAO. | GAO-18-472.

Figure 4. Nonprofit Volunteer Team Clearing Debris in Big Pine Key, Florida after Hurricane Irma.

[40] We began a review in the spring of 2018 of temporary sheltering challenges experienced during the 2017 disasters that will more fully examine the role of voluntary organizations in disasters.

The Federal Government Provided Support for Puerto Rico and the U.S. Virgin Islands, but Faced Multiple Challenges in Its Response

The federal government provided logistical support and conducted various preparedness actions in Puerto Rico and the U.S. Virgin Islands in advance of the hurricanes, as well as provided a high volume of response support to both territories after the hurricanes hit. Hurricane Maria made landfall on Puerto Rico as a category 4 hurricane, causing widespread infrastructural damages that left 3.7 million of the island's residents without electricity and 95 percent of cell towers out of service, and forced every airport and seaport on the territory to be closed. FEMA's response to Puerto Rico and the U.S. Virgin Islands included resources from dozens of federal agencies bringing unique capabilities to the response effort through 1,093 mission assignments totaling more than $5.5 billion, as of January 29, 2018. Forty-one percent of the mission assignments in Puerto Rico and the U.S. Virgin Islands were executed by the Department of Defense, resulting in more than $2 billion in obligations. However, FEMA faced challenges specific to Puerto Rico and the U.S. Virgin Islands that complicated its response efforts. As a result, FEMA has taken some action to incorporate lessons learned in preparation for the next hurricane season.

The Federal Government Provided Logistical Support and Conducted Various Preparedness Activities in Puerto Rico and the U.S. Virgin Islands in Advance of the Hurricanes

The federal government provided logistical support, deployed key incident management staff, and conducted numerous planning exercises prior to hurricanes Irma and Maria making landfall in Puerto Rico and the U.S. Virgin Islands in 2017. Specifically, FEMA shipped meals, delivered other commodities and activated contracts and mission assignments for additional federal support to the territories in advance of the hurricanes.

Further, FEMA had conducted various planning and response exercises from 2009 to 2017.

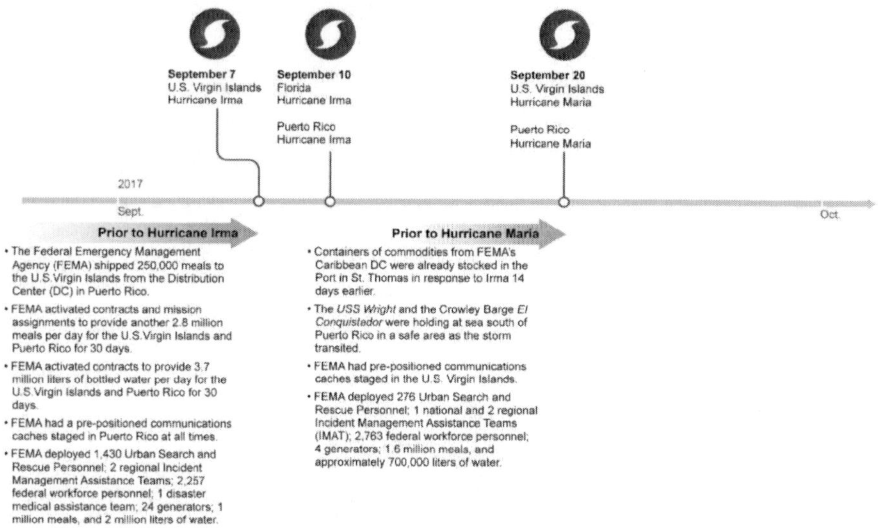

Source: Department of Homeland Security. | GAO-18-472.

Figure 5. The Federal Government's Logistical Support to Puerto Rico and the U.S. Virgin Islands In Advance of Hurricane Irma and Hurricane Maria.

These efforts included nine exercises that FEMA Region II—the FEMA regional office with oversight for both territories—conducted in 2017 for incident types with catastrophic impacts, such as tropical cyclones and tsunamis, in Puerto Rico and the U.S. Virgin Islands prior to hurricanes Irma and Maria. The agency also conducted an exercise in 2016 which, according to FEMA officials, allowed the regional and national IMATs the opportunity to integrate as one FEMA response team in preparation for a catastrophic disaster. According to FEMA officials, the exercise worked to enhance the capability and integration of the IMAT teams and the Regional Response Coordination Center—which coordinates federal response efforts during a disaster—to provide an effective response and resource support to Puerto Rico in the event of a catastrophic hurricane. DOD also conducted several events prior to the hurricanes that, according to FEMA officials, aided the response to hurricanes Irma and

Maria, such as the DOD Annual Joint Interagency Hurricane Terrain Walk and Exercise conducted in February 2017 in Puerto Rico.[41]

The Federal Government Provided a High Volume of Response Support to Puerto Rico and the U.S. Virgin Islands

According to FEMA's Office of Response and Recovery Assistant Administrator for Field Operations, FEMA's response to Puerto Rico was one of the largest recovery efforts in its history and included, among other things, bringing in food and supplies valued at approximately $1 billion; distributing food, commodities, and medicine via approximately 1,400 flights, which constituted the longest sustained air operations in U.S. disaster history; deploying 4,700 medical personnel; and utilizing USACE to install over 1,700 emergency electricity generators, as of May 2018, compared to 310 for the response to Hurricane Katrina. See Figures 6 and 7 for examples of support DOD provided to Puerto Rico after Hurricane Maria. According to the National Disaster Recovery Framework, local governments have the primary role in preparing for and managing the response and recovery of their communities, including leading pre-disaster recovery and mitigation planning efforts.[42] However, according to FEMA officials, FEMA essentially served as the first responder in the early response efforts in Puerto Rico. FEMA officials said that many services they provided—such as power restoration, debris removal, and commodity distribution—are typically provided by territorial or local governments.[43]

---

[41] According to FEMA, this event which was attended by multiple local and federal agencies, was focused on a hypothetical hurricane impacting the U.S. Virgin Islands and included a site survey of the Port of Ponce and local airport in the event of a Defense Support of Civil Authorities response.

[42] Department of Homeland Security, *National Disaster Recovery Framework*, Second Edition (June 2016).

[43] Debris removal services are generally provided by vendors that are contracted by local and territorial governments. While FEMA provides funding assistance for removal of debris from publicly and privately-owned lands and waters through the PA program, the respective state or local governmental entity generally has the responsibility to execute and manage debris removal operations.

In the U.S. Virgin Islands, recent disaster training and the pre-positioning of supplies due to the anticipated impact of Hurricane Irma facilitated the response efforts following Hurricane Maria, according to DOD officials. FEMA partnered well with local officials, according to the U.S. Virgin Islands Territorial Emergency Management Agency Director. According to FEMA's FCO for the U.S. Virgin Islands, the federal government deployed assets, including IMATs, urban search and rescue teams, and medical assistance teams. FEMA and USACE colocated with the U.S. Virgin Islands Territorial Emergency Management Agency in downtown St. Croix from the onset. In addition, due to the sequence of Hurricane Irma hitting the U.S. Virgin Islands immediately before Hurricane Maria, DOD already had personnel and resources (i.e., ships) deployed to the area, according to DOD officials, which enabled DOD to respond to Hurricane Maria faster than it otherwise would have.

Source: GAO. | GAO-18-472.

Figure 6. DOD Assigned USNS Comfort for Puerto Rico Response in November 2017 after Hurricane Maria.

Source: Federal Emergency Management Agency. | GAO-18-472.

Figure 7. Temporary Blue Roofs in Puerto Rico Installed by the U.S. Army Corps of Engineers after Hurricane Maria in February 2018.

Examples of Department of Defense's U.S. Army Corps of Engineers Post-Hurricane Maria Support

Puerto Rico

1,779 electricity generator installations 59,469 blue roof installations
4.12 million cubic yards of debris removed

The U.S. Virgin Islands

180 electricity generator installations 3,658 blue roof installations
More than 820,000 cubic yards of debris removed

Source: U.S. Army Corps of Engineers (May 2018) I GAO-18-472.

In both Puerto Rico and the U.S. Virgin Islands, DOD was also asked by FEMA to provide support that the department has not typically provided for prior hurricanes (e.g., air operations, mortuary affairs, and power grid restoration). According to DOD officials, active duty military personnel and reservists also provided life-sustaining commodities such as food and water. Additionally, USACE members provided services such as installing generators and tarp roofs. See appendix XI for a summary of DOD's role in the response effort and appendix III for more information on the full scale of federal support provided to Puerto Rico and the U. S. Virgin Islands, as well as the states affected by the 2017 hurricanes and wildfires.

FEMA Response Efforts Were Complicated By Factors Specific to Puerto Rico and the U.S. Virgin Islands

Major Factors that Affected Response to Hurricanes Irma and Maria

100 percent of Puerto Rico Electric Power Agency (PREPA) clients without electric power service

80 percent of all PREPA Infrastructure was destroyed

80 percent of Puerto Rico Aqueduct and Sewer Authority clients without water service

80 to 85 percent of communication towers not operational

Large numbers of roads with landslides and several dozen bridges collapsed

Maritime ports and airports closed for at least 5 days after Maria hit Puerto Rico

Satellite phones not working

Emergency Operations Plan not built for catastrophic levels nor 100 percent loss of communication

Ample fuel but not enough personnel and fuel tankers for transportation throughout the island

Source: Federal Emergency Management Agency. I GAO-18-472.

FEMA's response efforts in Puerto Rico and the U.S. Virgin Islands were complicated by a number of factors including (1) the remote distance of the territories, (2) limited local preparedness for a major hurricane, (3) outdated local infrastructure, (4) workforce capacity constraints, and (5) additional challenges in Puerto Rico.

These challenges were compounded by FEMA's previous deployment of personnel and assets to support the response for Hurricane Harvey in Texas and Hurricane Irma in Florida due to the unprecedented near-sequential disasters of 2017. As a result of lessons learned from these challenges, FEMA has taken a number of steps to plan for the next hurricane season.

Remote Distance of Territories

Given Puerto Rico's and the U.S. Virgin Islands' remote distance from the U.S. mainland, FEMA faced challenges in getting key personnel and resources to the territories before and after the hurricanes made landfall, and with distributing those resources to survivors. Both Puerto Rico and the U.S. Virgin Islands are located approximately 1,000 nautical miles from the U.S. mainland where personnel, equipment, and other key resources had to be moved from, as depicted in Figure 8.

Under typical disaster operations, responders are moved to the disaster response area via commercial travel options—or as in Texas and Florida, on roadways from nearby states, according to FEMA officials. However, limitations on air travel due to capacity constraints and power outages meant that FEMA had to coordinate and mobilize agency partners to provide chartered air transportation until commercial travel options resumed. Further, the destruction of major transportation routes made the deployment of these personnel and distribution of the commodities even more challenging according to FEMA officials. For example, in Puerto Rico, FEMA officials said food, commodities, and medicine had to be distributed by helicopter drop for several weeks because of landslides or destroyed bridges.

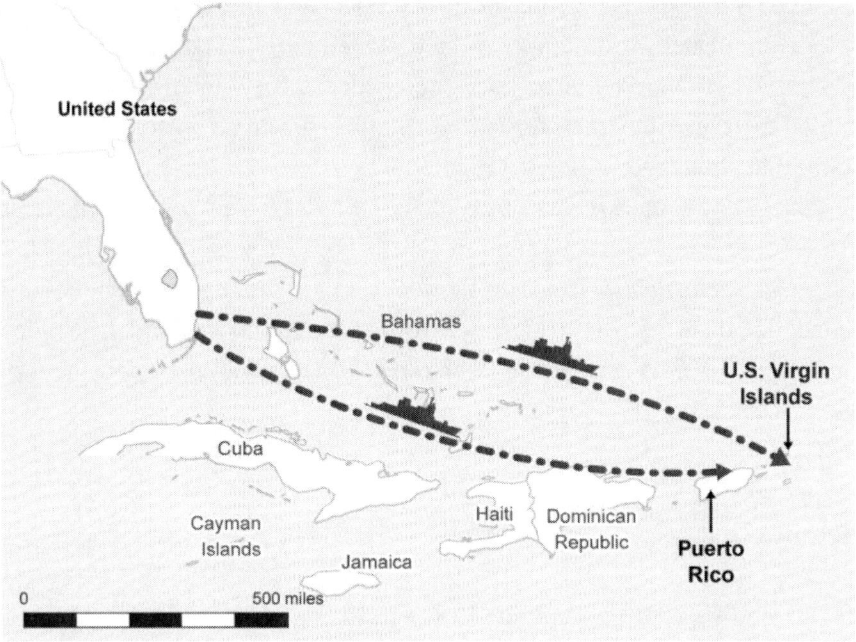

Source: Map Resources. | GAO-18-472.

Figure 8. Distance from U.S. Mainland to Puerto Rico and the U.S. Virgin Islands.

Limited Local Preparedness

Puerto Rico and the U.S. Virgin Islands had engaged in disaster preparedness exercises prior to Hurricane Maria; however, neither had recently experienced nor stockpiled the resources necessary for a hurricane of that magnitude. For example, Puerto Rico officials said their emergency plans allowed the local government to respond effectively to Hurricane Irma (e.g., evacuating residents, purchasing food, and securing their homes). However, their plans were insufficient for the magnitude of Hurricane Maria which made landfall 2 weeks later. Specifically, Puerto Rico officials had not considered that a hurricane would cause a loss of power for as long as Hurricane Maria did.

Hurricane Maria was the strongest hurricane to make landfall in Puerto Rico since a Category 5 hurricane in 1928, according to the National

Oceanic and Atmospheric Administration.[44] Puerto Rico officials explained that local preparation for a Category 5 hurricane is limited by physical space and financial resources needed to stockpile necessary supplies to respond to a hurricane of that magnitude and also because such hurricanes occur infrequently. According to FEMA officials, FEMA took on a more active role in the response to Hurricane Maria due to preparedness challenges in Puerto Rico.

In the U.S. Virgin Islands, the local government had conducted preparedness exercises and local officials had a grasp of the emergency management process, according to DOD officials.[45] According to FEMA officials, these preparedness exercises were for tropical cyclones and other incident types such as an earthquake and a tsunami which have catastrophic impacts similar to hurricanes Irma and Maria. However, according to FEMA's *2017 Hurricane Season FEMA After-Action Report*, FEMA could have better leveraged information from these and other prior exercises in the Caribbean, including a 2011 exercise after-action report for Puerto Rico which indicated that the territory would require extensive federal support in moving commodities, including from the mainland to the territory, and to distribution points throughout the territory. In contrast, FEMA's leverage of information from prior exercises in Florida proved to be critical in that state's ability to efficiently execute mutual aid agreements in response to Hurricane Irma.

FEMA officials said they have encouraged both Puerto Rico and the U.S. Virgin Islands to develop planning timelines and formal Emergency Management Assistance Compacts (EMAC)—mutual aid agreements that allow states to support one another during a disaster response—which play a critical role in managing risk to communities and infrastructure.

---

[44] National Oceanic and Atmospheric Administration (NOAA) measures hurricanes on a scale from 1 to 5 with a Category 1 being the least intense and a Category 5 being the most intense. NOAA defines a Category 4 hurricane as one with winds 130-156 miles per hour and Category 5 with winds above 157 miles per hour. Hurricane Maria made landfall on the U.S. Virgin Islands as a Category 5 hurricane and Puerto Rico as a high end Category 4 hurricane.

[45] Exercises are a useful tool for jurisdictions to identify emergency preparedness capability strengths and shortfalls which can be used to inform future preparedness efforts and response operations.

Although an EMAC supported requests for assistance to Puerto Rico, getting interstate mutual aid and assistance through the EMAC process was more difficult because there are no other states or territories adjacent to Puerto Rico and the U.S. Virgin Islands, according to DOD officials.[46]

Outdated Local Infrastructure

Source: GAO. | GAO-18.472.

Figure 9. Damaged Power Pole in the U.S. Virgin Islands after Hurricane Maria in November 2017.

[46] We are currently conducting a separate review on Puerto Rico and the U.S. Virgin Islands' disaster recovery plans, which we plan on issuing in spring 2019.

Hurricane Maria devastated the already fragile and outdated infrastructure in Puerto Rico and the U.S. Virgin Islands, which complicated response efforts according to the FEMA Administrator and Puerto Rico officials.

Specifically, Hurricane Maria crippled the power grid, communication systems, and transportation infrastructure throughout both territories, hindering communication and delaying emergency response activities. See Figures 9 and 10 for photographs of damage sustained in the U.S. Virgin Islands and Puerto Rico. The damage was more extensive in Puerto Rico. Three months after Hurricane Maria hit, Puerto Rico had 65 percent of its power restored while the U.S. Virgin Islands had closer to 90 percent, according to FEMA officials. The U.S. Virgin Islands was quicker to restore power in part due to previous infrastructure investments and mitigation efforts implemented by the government, according to FEMA officials.

Source: Federal Emergency Management Agency. | GAO-18-472.

Figure 10. Damaged Power Lines and Satellite Dish in Puerto Rico after Hurricane Maria in November 2017.

According to Puerto Rico officials, much of the territory's infrastructure (e.g., roads, sewage systems, and bridges) is more than 50 years old. As a result, some replacement parts were no longer available and had to be specially manufactured, delaying power restoration. Moreover, many of the power lines connecting the large power stations in southern Puerto Rico through the mountains to the north—where the majority of the island's residents reside—were destroyed by the storm according to officials from the Puerto Rico Electric Power Authority (PREPA).

According to FEMA and Puerto Rico Aqueduct and Sewer Authority officials, Puerto Rico had limited access to federal funds to renew and replace infrastructure prior to hurricanes Irma and Maria due to its outstanding public debt.[47] Specifically, PREPA officials told us that their ability to prepare for the hurricane season was impacted by Puerto Rico's financial situation due to vendor concerns with reimbursement for services and goods.[48] Moreover, previous hazard mitigation efforts to strengthen the island's infrastructure were not enough to withstand the force of Hurricane Maria, according to FEMA officials. For example, Puerto Rico's construction codes called for buildings to withstand winds of 145 miles per hour, but Hurricane Maria's winds exceeded 175 miles per hour causing massive wind and flood damage to housing and office space, including the Puerto Rico Emergency Management Agency office, further delaying hurricane response efforts. We also provide more details about both

---

[47] Puerto Rico has roughly $70 billion in outstanding debt and $50 billion in unfunded pension liabilities, and since August 2015 has defaulted on over $1.5 billion in debt. The effects of hurricanes Irma and Maria will further affect Puerto Rico's ability to repay its debt, as well as its overall economic condition. For more on Puerto Rico's public debt see GAO, *Puerto Rico: Factors Contributing to the Debt Crisis and Potential Federal Actions to Address Them,* GAO-18-387 (Washington, D.C.: May 9, 2018).

[48] Congress passed the Puerto Rico Oversight, Management, and Economic Stability Act (PROMESA) in June 2016. See Pub. L. No. 114-187, 130 Stat. 549 (2016). PROMESA temporarily prevented creditors from suing Puerto Rico over missed debt payments. PROMESA established a Financial Management and Oversight Board with broad powers of budgetary and financial control over Puerto Rico. In addition, it created procedures for adjusting debts accumulated by the Puerto Rico government and its component units. On May 3, 2017, after the termination of the original stay preventing creditors from suing the territory, the board filed a petition under Title III of PROMESA beginning a broad-based debt restructuring process.

territories' challenges as well as goals for incorporating resilience after the 2017 disasters in appendix X.[49]

Workforce Capacity Constraints

Given that FEMA was confronted with concurrently responding to four large and complex disasters, this exacerbated FEMA workforce capacity constraints and training deficits for deployed workers according to FEMA officials. This was especially true for Hurricane Maria, which was the last of the three major hurricanes in 2017.[50] FEMA leadership and the FCO in the U.S. Virgin Islands said that some of FEMA's disaster staff deployed to Puerto Rico and the U.S. Virgin Islands were not physically able to handle the extreme or austere environment of the territories, which detracted from mission needs. According to FEMA officials, the physical fitness of staff could be assessed prior to deploying staff.

Similar to Texas and Florida, federal and local response agencies were colocated at the JFO in Puerto Rico and the U.S. Virgin Islands to facilitate coordination and response. However, according to the IMAT leader in the U.S. Virgin Islands, the U.S. Virgin Islands IMAT team needs more exercises alongside the national IMAT teams in order to build response skills and strong relationships.

In addition, with much of the housing destroyed in both Puerto Rico and the U.S. Virgin Islands, locating accommodations for the nearly 15,000 federal government employees—including military personnel—deployed to assist in response activities became a major challenge,

---

[49] We are also currently conducting a more detailed review of the power grid restoration and resilience efforts in Puerto Rico and the U.S. Virgin Islands, as well as USACE power restoration capacity and efforts in Puerto Rico, which we plan to issue in spring 2019.

[50] According to FEMA officials, FEMA employed use of a force package for the first time to Hurricane Harvey. A force package is a set of force modules with time-phased deployment dates and each force module within a package provides a capability for disaster response. The force package provides the personnel to support an event based on the incident size and scope. Field leadership is provided with an initial deployment of time-phased personnel (supervisors and management personnel arriving first) allowing them to better manage further deployments and staffing requests on the margins or as the operation changes over time.

according to FEMA officials. We discuss more of the workforce challenges FEMA faced in the wake of the 2017 disasters later in this chapter. See appendix V for a summary of FEMA's workforce capacity and related challenges during the 2017 disaster response.[51]

Other Challenges in Puerto Rico

Prior to Hurricane Maria's landfall, FEMA identified some factors unique to Puerto Rico that could affect response efforts. Specifically, FEMA's Region II Hurricane Annex for Puerto Rico and the U.S. Virgin Islands, dated June 2014, outlines the territories' emergency response policies, procedures, and responsibilities, including communication, transportation, and mass evacuation in the event of a disaster.[52] According to FEMA's Region II Hurricane Annex, information in Puerto Rico must be conveyed in Spanish—which is the main spoken and written language. However, in the aftermath of Hurricane Maria, FEMA did not have enough bilingual employees to communicate with local residents or translate documents. According to FEMA officials, this resulted in further delays while staff were reshuffled from other disasters to Puerto Rico. In addition, unlike in the continental United States where individuals mostly apply for IA at disaster recovery centers or online, in Puerto Rico, officials needed to conduct more door-to-door visits to reach disaster survivors and conduct assessments, according to FEMA officials. Locating addresses and individuals was challenging, according to FEMA officials, because many affected areas did not have posted addresses, many individuals use nicknames instead of their given names, and often several families were located on a single property. Additionally, we are conducting work on how

[51] We are currently conducting a separate review focused on FEMA's workforce management challenges.

[52] According to FEMA, the purpose of FEMA's Region II Hurricane Annex for Puerto Rico and the U.S. Virgin Islands is to support the expedited jurisdictional response to tropical and subtropical systems, including catastrophic hurricanes, as well as tropical depressions, tropical storms, and hurricanes, and their secondary and cascading impacts on locations in Puerto Rico and the U.S. Virgin Islands. FEMA Region II Hurricane Annex for Puerto Rico & the U.S. Virgin Islands. June 1, 2014.

states and territories account for disaster-related deaths and injuries, and the impact on disaster assistance.

FEMA Has Taken Some Action to Incorporate Lessons Learned in Preparation for the Next Hurricane Season

Following the 2017 hurricane season, FEMA has taken a number of steps to help prepare for the 2018 season. For example, in July 2018, FEMA issued an after-action review of the agency's preparations for, immediate response to, and initial recovery operations for hurricanes Harvey, Irma, and Maria. The after-action report identifies 18 key findings across five focus areas and offers targeted, agency-wide recommendations for improvements as well as broader lessons for the emergency management community.[53] FEMA officials reported that they have already taken several actions in preparation for the 2018 hurricane season. These include:

- updating hurricane plans, annexes, and procedures for all U.S. states and territories;
- improving staff skills and readiness including by creating a Standard Operating Procedure for a central location for equipping and training staff prior to disaster deployments;
- improving logistics operations such as by increasing disaster supplies for all U.S. territories for items including meals, water, tarps; and generators, specifically for Puerto Rico and the U.S. Virgin Islands. FEMA is also adding 360 new emergency generators to its inventory and has pre-positioned 630 generators in the Caribbean for the 2018 hurricane season; and

---

[53] The five focus areas of the 2017 Hurricane Season FEMA After-Action Report include: (1) scaling a response for concurrent complex incidents; (2) staffing for concurrent complex incidents; (3) sustaining whole community logistics operations; (4) responding during long-term Infrastructure outages; and (5) mass care to initial housing operations.

- updating communications systems from land-based radios to satellite-based technology; refining tactical and long-haul communications, from land mobile radios to satellite communications.

In addition to these efforts, FEMA conducted a National Level Exercise to assess and enhance its response and initial recovery capability during the first 2 weeks of May 2018, which focused on issues identified in its 2017 hurricane season after-action review.[54] The findings and recommendations of the after-action report influenced the development of FEMA's 2018-2022 Strategic Plan, released in March 2018, according to FEMA officials.

While FEMA described actions it has taken in response to the 2017 disasters, it was too soon to assess the adequacy of these actions as part of this review and whether the actions will have the intended impact.[55]

THE 2017 HURRICANES AND WILDFIRES HIGHLIGHTED CHALLENGES FEMA FACES IN MEETING ITS RESPONSE AND RECOVERY MISSIONS

The 2017 hurricanes and wildfires reaffirmed the existence of some longstanding response and recovery challenges, but also highlighted several new challenges related to (1) the near-sequential timing of the

---

[54] According to FEMA, the National Level Exercise as a whole is a large exercise series running from January through the summer of 2018, which includes seminars, workshops, and tabletop exercises, as well as a functional exercise in May focused on thematic areas identified from ongoing real-world continuous improvement efforts. The 2018 National Level Exercise examined the ability of all levels of government, private industry, and nongovernmental organizations to protect against, respond to, and recover from a major Mid-Atlantic hurricane, and allowed the whole community to examine lessons observed following the storms of the 2017 Atlantic hurricane season.

[55] We are conducting a comprehensive review of the federal government's national preparedness capabilities, training, and funding to assist communities in responding to and recovering from major disasters, which we plan to issue in the summer of 2019.

disasters, (2) housing assistance, (3) workforce management, and (4) public assistance.

The Timing of the Four Near-Sequential Major Disasters and Unique Circumstances in Each Location Presented Challenges for Debris Removal in 2017

Debris removal is an important first step in the disaster recovery process, allowing communities to expedite the recovery process by restoring accessibility to public services and space, while ensuring public health and safety in the aftermath of a disaster. The PA Program provides funding assistance for the removal of debris and wreckage from publicly and privately-owned lands and waters resulting from a major disaster, when such removal is in the public interest.[56] For example, debris removal is in the public interest if it would eliminate an immediate threat to lives, public health and safety, or property. In addition to funding, FEMA can provide a range of assistance and guidance to help PA applicants mitigate associated difficulties with debris removal. For example, FEMA can provide Direct Federal Assistance for debris removal if the event exceeds state and local capability and if the recipient requests such support. When local jurisdictions do not have the capacity for debris removal, FEMA can assign the USACE or other federal agencies to conduct the removal, or other agencies may support these efforts through various authorities. However, the respective state or local governmental entity has the responsibility to execute and manage debris removal operations.

Officials in Texas, Florida, Puerto Rico, and California reported challenges with debris removal operations following each disaster in 2017, for example, clearing trees and other vegetative debris, as well as residential, commercial, and construction goods, including hazardous materials. The officials said these challenges arose from a shortage of

---

[56] See 42 U.S.C. § 5173; 44 C.F.R. 206.224.

debris removal contractors, inadequate debris contract provisions, and disputes over responsibility for marine debris removal.

Shortage of Debris Removal Contractors

Some Texas and Florida jurisdictions experienced challenges with the availability of debris removal contractors, given the large demand throughout the regions, and in some cases, even when contracts were in place, contractors did not honor them. For example, according to the FEMA FCO in Florida, contractors are primarily located in the Gulf Coast area and serve a national market. Because many debris contractors were already engaged in Texas, the contractors were not available to provide services in Florida even if they had prior contracts in place. Additionally, some officials from local jurisdictions reported that they had existing contracts for debris removal that had been in place up to 5 years ago. However, because newer contracts were offering more money per cubic yard, some vendors, despite being under contract in a given jurisdiction, prioritized work in another jurisdiction that offered higher rates, according to local officials. Florida officials said that in such cases there was not much they could do except to urge fulfillment of the contract terms.

According to local officials in one Texas jurisdiction, their debris removal contracts are awarded based on price, with preference given to the lowest bids, which can make it difficult for them to get debris removal services in a timely manner in times of high demand as happened in 2017, because contractors inevitably prioritize jurisdictions that pay more. In this case, the officials asked FEMA to issue one regional contract at a higher bid price to ensure all jurisdictions had the opportunity to access debris removal services. According to FEMA's Policy Branch Chief, the magnitude of debris across very large geographic areas, the lack of proximity of debris removal contractors to certain areas, and the limited number of debris removal contractors providing such services resulted in much of the difficulty related to debris removal experienced after the 2017 hurricanes. It also resulted in the inability of multiple states and municipalities to quickly procure a debris removal contactor at a reasonable rate. See Figure 11 for an example of a debris pile awaiting pick up in a Texas residential neighborhood following Hurricane Harvey.

Source: GAO. | GAO-18-472.

Figure 11. Residential Debris Awaiting Pick Up in Texas Following Historic Flooding Caused by Hurricane Harvey in 2017.

In addition to the funding assistance that FEMA can provide for debris removal operations, FEMA officials said that the agency also has an online debris contractor registry to assist PA applicants in identifying and contacting contractor resources. Additionally, through its regional counsel at Joint Field Offices and through its Procurement Disaster Assistance Team, FEMA proactively engages with communities to provide technical assistance and guidance to PA applicants for debris removal contracting. However, FEMA officials stated that greater readiness and preparedness at the state level may be able to address some of the issues states and municipalities experienced with debris removal in 2017. FEMA PA officials told us that they are exploring potential opportunities to provide additional guidance to applicants on federal procurement rules to assist preparedness efforts. FEMA is also considering the potential utility of new debris removal estimating methodologies and technologies to serve as a basis for providing debris removal funding based on estimates as opposed to actual costs under the PA alternative procedures.

Inadequate Debris Contract Provisions

In California, the state's primary debris contractor did not have the capacity to handle all of the debris removal after the 2017 wildfires. Therefore, California OES coordinated with FEMA, which directed USACE to provide debris removal services. According to the California OES director, the state experienced challenges with meeting the requirements for a federal contract and the lack of flexibility resulted in delays with the debris removal process. Moreover, according to FEMA Region IX officials in California, because wildfires create ash debris, asbestos, and other toxic chemicals, debris removal contracts must have specific provisions to address fire debris. However, the region's advance contracts with USACE for debris removal did not have a task order for fire debris because this was the first time FEMA requested USACE support for fire debris removal. This caused issues in the debris removal process.

Disputes over Responsibility for Marine Debris Removal

In the Florida Keys, determining responsibility for marine debris removal from the canals was a challenge after Hurricane Irma, according to the Monroe County emergency management director. County officials estimated that there were between 1,300 and 1,800 sunken or derelict boats for which neither the private owners who use the water canals nor the Coast Guard nor the county would claim responsibility for debris removal.[57] In the absence of available contractors, some jurisdictions opted to use their own staff and equipment to remove debris. However, these jurisdictions faced fiscal challenges because of FEMA delays processing reimbursements.

Source: GAO. | GAO-18-472.

Figure 12. Marine Debris in Florida Keys Canal Following Hurricane Irma in 2017.

Responsibility for marine debris removal has been a longstanding issue with parties reaching different solutions at different times, according to Monroe County officials in Florida. See Figure 12 for an example of

[57] We reported in 2017 that abandoned and derelict vessels can block navigable U.S. waterways and pose threats to the environment, public health and safety, as fuel and hazardous material can leak into the water as the vessels deteriorate. See GAO, *Maritime Environment: Federal and State Actions, Expenditures and Challenges to Addressing Abandoned and Derelict Vessels,* GAO-17-202 (Washington, D.C.: Mar. 28, 2017). Federal agencies respond to abandoned and derelict vessels in accordance with federal law, interagency agreements, and funding availability. Federal laws and the National Contingency Plan—the government's blueprint for responding to oil and hazardous substance releases—establish federal agency roles for leading a response to an abandoned and derelict vessels-related incident based on various factors, such as the type of abandoned and derelict vessels threat posed and its location. Interagency agreements have also helped to guide federal abandoned and derelict vessels response efforts.

marine debris in the Florida Keys, responsibility for the removal of which was in dispute at the time of our November 2017 visit there.

According to FEMA officials, the agency offers PA applicants financial incentives through its PA Alternative Procedures Debris Removal Pilot Program to establish Debris Management Plans, written procedures, and guidance for managing debris removal in an expeditious, efficient, and environmentally sound manner. FEMA also provides PA applicants with a job aid to assist applicants in developing the Debris Management Plans, according to FEMA's Policy Branch Chief. Applicants that develop and have the plan in place prior to a declaration and meet other program requirements can take advantage of a one-time 2 percent federal cost share increase on debris removal operations for 90-days following the 1st day of the incident period. Additionally, FEMA reviews the plan and informs applicants of deficiencies and how such deficiencies could be remedied, which could result in faster processing of their project and therefore reimbursement, if action is taken to address the deficiencies.[58]

FEMA, States, and Territories Are Using a Variety of Existing and New Short-Term Housing Options for Disaster Survivors, but Challenges Persist

According to DHS's 2017 National Preparedness Report, providing effective and affordable short-term housing for disaster survivors has been a longstanding and continuing challenge.[59] For example, according to the report, many states and territories expect the federal government to take on the responsibility of addressing housing gaps, as states often face shortages in effective housing options following a large-scale disaster. Moreover, short-term housing options that might work in one location may not be suitable for another for various reasons, according to the report, hence the

[58] We have initiated two reviews examining advance and post-disaster contracting, which will include contracts that were used for debris removal, and the DHS Office of Inspector General plans to conduct future reviews of disaster debris removal contracts.

[59] Department of Homeland Security, *National Preparedness Report* (Washington, D.C.: Aug. 28, 2017).

need for FEMA to be flexible in its implementation of potential housing solutions. The FEMA Administrator has also highlighted these challenges, noting that state and local officials, not FEMA, are in the best position to determine the necessary housing options for their citizens, with support from the federal government. The Administrator cited the various federal housing models states used after the 2017 hurricanes and wildfires. These models included Direct Lease, Multifamily Lease and Repair, Manufactured Housing Units and Recreational Vehicles, Permanent Housing Construction Repair Program, Sheltering and Temporary Essential Power, and Transitional Sheltering Assistance. According to FEMA officials, these alternative housing models were identified as a result of lessons learned from recent major disasters such as the 2016 flooding in Louisiana and Hurricane Matthew as well as insights leveraged from a housing summit the agency held in February 2017—FEMA's Housing Assistance Initiative.[60]

- Direct Lease. FEMA provides temporary housing units directly to survivors when rental resources are unavailable. FEMA and the state of Florida are prioritizing use of this approach for housing recovery in the state, according to FEMA officials.
- Multifamily Lease and Repair. According to FEMA, FEMA repairs existing multi-family housing units, such as apartments, to use as temporary housing for eligible applicants who are unable to use Rental Assistance—a grant in the form of a check to enable survivors, both homeowners and renters, to rent temporary replacement housing—due to a lack of available resources.
- Manufactured Housing Units and Recreational Vehicles. These are manufactured homes or other readily fabricated dwellings owned by FEMA and provided to eligible applicants for use as temporary housing for a limited time. Recreational vehicles have been

[60] FEMA's Housing Assistance Initiative integrated the efforts of 13 working groups and ongoing housing initiatives to develop an implementation plan that outlines short-term and long-term changes to how FEMA provides disaster housing and that outlines a new strategic vision that engages states, other federal agencies, and industry to grow national disaster housing capabilities.

approved for use in response to hurricanes Harvey and Irma. This form of assistance is being implemented in Texas and Florida.

- Permanent Housing Construction. FEMA may provide financial assistance or direct assistance to individuals and households in insular areas outside the continental United States, or in other locations where no alternative housing resources are available and where temporary housing assistance is unavailable, infeasible, or not cost-effective. Assistance may be authorized for direct repairs or new home construction; however, FEMA has only authorized Permanent Housing Construction. This is one element of the new approach to housing recovery utilized in Texas, Puerto Rico, and the U.S. Virgin Islands, according to FEMA officials.
- Sheltering and Temporary Essential Power. FEMA uses Sheltering and Temporary Essential Power to assist state, territorial and tribal governments in performing work and services essential to saving lives, protecting public health and safety, and protecting property to enable survivors to shelter at home, according to FEMA officials.
- Transitional Sheltering Assistance. FEMA may provide Transitional Sheltering Assistance services to applicants who are unable to return to their pre-disaster primary residence because their home is either uninhabitable or inaccessible. The goal of this program is to reduce the number of disaster survivors in congregate shelters by transitioning survivors into short-term accommodations through direct payments to lodging providers, such as hotels. Puerto Rico used this approach despite initial concerns that this would have a negative effect on migration away from the island territory, according to FEMA officials.

See Table 1 for the number of eligible and approved applicants for each of these housing and sheltering assistance options for each disaster location, and appendix VI for more information on how and where these sheltering and housing approaches were implemented in response to the 2017 disasters.

Table 1. Number of Registrations and Approved Applicants for Each Type of Housing and Sheltering Assistance Provided by Disaster Location, as of June 22, 2018

Disaster	Total Registrations	Multifamily Lease Repair	Manufactured Housing Units and Recreational Vehicles	Permanent Housing Construction Repair Program	Direct Lease	Sheltering and Temporary Essential Power Program	Transitional Sheltering Assistance
Hurricane Harvey in Texas	895,528	0	2,848	247	131	15,578	53,894
Hurricane Irma in Florida	2,644,403	0	257	0	63	129	26,633
Hurricane Irma and Hurricane Maria in Puerto Rico	1,138,444	16	0	33	237	33,016	6,907
Hurricane Irma and Hurricane Maria in U.S. Virgin Islands	39,415	0	0	0	61	1,920	0
California wildfires	25,425	0	154	0	94	0	618
Total	4,743,215	16	3,259	280	586	50,643	88,052

Source: Federal Emergency Management Agency I GAO-18-472.

Note: Total registrations represent the total number of survivors considered for disaster assistance at each of the disaster locations specified. However, the number of registrations for the Sheltering and Temporary Essential Power Program may be different because the Sheltering and Temporary Essential Power Program is separate from FEMA's Individuals and Households Program and is not included in the registration process. Disaster survivors interested in the Sheltering and Temporary Essential Power Program must first apply for FEMA's Individuals and Households Program assistance and then apply directly to the State, Territorial, or Tribal entity administering the Sheltering and Temporary Essential Power Program.

These alternative approaches for housing recovery—primarily funded through FEMA's IA program—reflect the agency's efforts to support recovery efforts that are responsive to local needs and available resources, and that address previously identified challenges facing their programs. However, despite successes, in each of the disaster-affected areas, officials noted complex and ongoing housing concerns, as discussed below.

Texas

In September 2017, FEMA entered into an agreement with the Texas General Land Office to provide for housing recovery, marking the first time the agency has coordinated with a nonfederal agency to provide this housing service, according to FEMA officials. FEMA estimates these costs will reach approximately $1 billion. State officials in Texas plan to implement the new housing model to manage the delivery of direct housing to more than 6,600 applicants.[61] However, the officials cited staffing shortfalls at the state level, and information sharing challenges among FEMA, state, and local officials, which may result in delays in granting housing relief to applicants.

Florida

Housing shortages already existed in Florida prior to Hurricane Irma, and locating adequate housing subsequent to the disasters has been a major challenge, according to state and local officials. In Florida, officials told us that that zoning laws restrict the number of housing units based on the capacity for evacuation from the area. As a result, state officials said they preferred to utilize the direct lease option for the first time to leverage the high volume of vacation rentals, particularly in Lee, Collier, and Monroe

[61] Under the new model developed immediately after Hurricane Harvey, the General Land Office—a Texas state agency—acts as the program administrator, while local government officials each individually run the program, which provides a range of housing options to applicants, including permanent housing construction and direct lease. According to FEMA officials, after-action reviews which highlighted challenges experienced in the Baton Rouge flooding of 2017 drove this model of contracting with the state to provide housing assistance. As part of that review, FEMA captured lessons learned from the flooding event that recommended, among other changes, giving grant authority to enable states to develop capability in advance of a disaster.

Counties. However, local officials said in December 2017 that housing shortages existed in areas of the Florida Keys, and that there were concerns about the timeliness of providing housing units to disaster survivors.

U.S. Virgin Islands

Significant wind damage from Hurricane Irma followed by significant water damage from Hurricane Maria severely depleted the housing available for disaster survivors. Local officials stated that they are addressing the shortage and challenges through a combination of direct temporary housing—direct lease and multifamily lease and repair, and potentially permanent housing construction repairs. However, FEMA officials noted challenges in providing housing to disaster survivors and local officials due to staff shortages for inspections and inexperienced FEMA staff who are not prepared to support the housing mission. According to FEMA's Housing Inspection Services officials, to expedite availability of trained inspectors, FEMA pursued and received a waiver of the standard background check process for inspectors during recruitment efforts following Hurricane Harvey and throughout the 2017 hurricane season. The process, which normally takes about 2 weeks, was abbreviated to approximately 2 to 3 days. It also initiated innovations for the inspection process that included streamlining the scope of field inspection to more readily allow third party representatives to meet with the inspector and a self-declaration process as a last resort to verify either occupancy or ownership.

Puerto Rico

FEMA and territorial officials are offering multiple programs to address the unique challenges to long-term housing for survivors on the island. For example, the governor considered the shortage of hotel space available on the island and ultimately decided to request Transitional Sheltering Assistance, despite concerns that it would encourage middle class, professional residents to leave the island, according to FEMA officials.[62] The governor also requested transportation assistance for the first time to assist people who wish to relocate, including those in nursing homes and hospitals, FEMA officials added.[63]

California

Even before the 2017 disasters, critical housing shortages existed in California. According to FEMA officials, Sonoma County, California had 144 available housing units prior to the wildfires, which burned 5,098 homes in Sonoma and another 5,031 homes in Mendocino County, creating a large housing deficit. Locating adequate housing subsequent to the disasters has been a major challenge, according to California state officials. These officials reported that local ordinances have compounded the complexity of finding housing solutions, and local officials navigate alterations to rules for temporary situations (e.g., prohibiting mobile homes in certain areas).

Federal, state, and local officials formed housing task forces to inform the state and FEMA of local challenges as well as opportunities to provide shelter. According to state officials, the pre-existing relationship with FEMA along with California's more centralized emergency management decision-making structure has facilitated this joint decision-making approach. While this approach has enabled the state to meet its most pressing short-term housing needs, according to FEMA officials, the state faces other challenges in the long term. For example, FEMA officials

---

[62] See 42 U.S.C. § 5189c (authorizing the provisions of transportation assistance to relocate individuals displaced from their predisaster primary residences as a result of a declared major disaster or to return an individual or household to their pre-disaster primary residence or alternative location). According to FEMA, the 2017 disaster season was the largest implementation of Transitional Sheltering Assistance in the agency's history. In addition to Puerto Rico, this type of assistance was applied in Texas, Florida, and California in response to the 2017 hurricanes and wildfires.

[63] According to FEMA, a mission assignment was approved for National Disaster Medical System to provide a medical evacuation of dialysis patients from both the US Virgin Islands and Puerto Rico to the Continental United States. Officials said that transportation assistance was not approved or used to assist Puerto Rico survivors in leaving the island. Transportation assistance was approved on May 3, 2018, however, to assist survivors that were checked into Transitional Sheltering Assistance (TSA) hotels within the Continental United States in returning to Puerto Rico. This assistance was provided from May 3, 2018 through August 30, 2018.

estimated that Direct Temporary Housing, which should be occupied by disaster survivors no more than 18 months, will be needed for a period of at least 36 months by survivors of the wildfires since that is how long it is estimated that survivors will take to rebuild.[64]

Although FEMA and state officials acknowledge the potential of the alternative approaches to meet local needs, early implementation challenges raise concerns about the effectiveness and management of these approaches for long-term recovery. We describe these challenges and approaches for each location further in appendix VI.[65]

FEMA Faced Workforce Challenges With Deploying Enough Personnel, Providing Training, and Retaining Staff

FEMA's workforce allocations and plans were overwhelmed by the 2017 disaster response needs and long-standing workforce challenges we have identified in prior work were exacerbated by the need to provide a concurrent response to the disasters. Based on its internal workforce analyses, FEMA faced a staff shortage of more than 30 percent as of September 1, 2017.[66] Among other things, the Post-Katrina Act required FEMA to develop a strategic human capital plan that includes an assessment of the critical skills and competencies of FEMA's workforce and provide an action plan that includes workforce planning strategies and program objectives to train employees.[67]

---

64 According to FEMA's Individuals and Households Program Unified Guidance, FEMA may extend Direct Temporary Housing Assistance beyond the 18-month period of assistance when the affected state, territorial, or tribal government requests an extension in writing. See also 42 U.S.C. § 5174(c)(1)(B)(iii).

65 We are initiating a comprehensive review of FEMA's Individual Assistance Program—including FEMA housing assistance—that will examine challenges and lessons learned from the 2017 disasters, which we plan to issue in 2019.

66 FEMA's workforce analysis only represented FEMA employees and did not include local hires, FEMA Corps, Surge Capacity Force, contractors or employees from other federal agencies.

67 See 5 U.S.C. § 10102.

We have previously found that FEMA has faced challenges developing workforce strategies and ensuring adequate training that affects FEMA's ability to ensure workforce capacity to respond to and recover from large-scale disasters.[68] In our prior reports, we have recommended, among other things, that FEMA incorporate certain principles, including goals and performance measures, into its workforce planning and training efforts. FEMA concurred, and has taken steps to address some of our recommendations. However, FEMA officials said that the agency does not expect to address all of our recommendations until 2020.

Deploying Personnel

Prior to the 2017 disasters, one of FEMA's strategic goals was to maintain a FEMA disaster workforce that is capable of responding to two concurrent large-scale disasters. Following the 2017 disasters, FEMA officials told us that their experience responding to four near-simultaneous disasters made them realize that they will need to continue to improve their workforce planning to be prepared to simultaneously support multiple disasters. FEMA relies on both permanent and disaster-related temporary employees to respond to presidentially-declared disasters. See Figure 13 for the total federal disaster workforce, including the FEMA disaster workforce, deployed to the various disasters in 2017.

According to FEMA officials, the near-sequential disasters in 2017 required FEMA management to redeploy response personnel from one disaster to the next. For example, FEMA initially deployed a national IMAT team of 32 personnel to Texas following Hurricane Harvey, after hurricanes Irma and Maria hit, FEMA replaced this team with a regional IMAT team of 12 personnel, according to Harris County officials in Texas. Also, according to the FEMA deputy FCO in California, FEMA had already deployed the majority of its workforce to support the hurricanes when the wildfires began and so there was some delay in initially deploying an adequate number of staff to support the wildfires response.

---

[68] GAO-12-487.

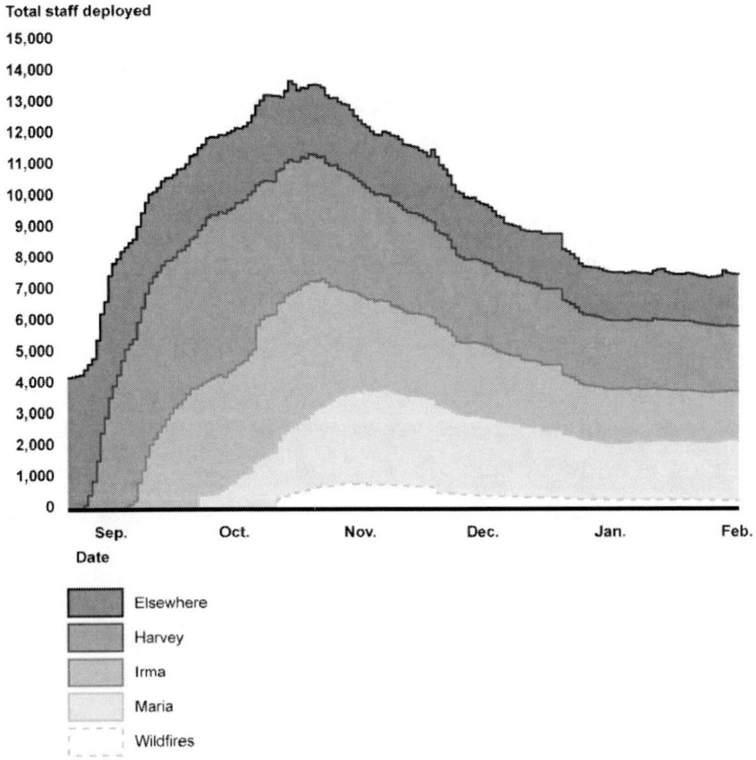

Source: GAO analysis of Department of Homeland Security data. | GAO-18-472.

Note: Data include Federal Emergency Management Agency (FEMA) workforce and non-FEMA employees that FEMA can deploy during a disaster response such as the Surge Capacity Force, FEMA Corps and contractors. Data do not represent local hires or employee types such as permanent part-time, temporary part-time, and temporary incident employees.

Figure 13. Federal Disaster Workforce Deployed in Response to the 2017 Disasters in September 2017 through February 2018.

According to FEMA Region IX officials responsible for responding to the California wildfires, FEMA classified the hurricanes as the most complex level of disaster, while FEMA initially classified the wildfires at a lower level of complexity.[69] The officials added that as a result, it was

[69] FEMA defines disasters by incident types. A Type 1 event is the most complex, requiring national resources to safely and effectively manage and operate. Other levels of disaster do not call for the same level of national resources.

difficult for the region to get skilled staff into positions that were crucial for the response, so the region accepted employees who could only deploy for a week or two when FEMA would normally deploy them for 30 days. According to the officials, the region had retained some staff in California and was able to use contractors to backfill positions in cases where FEMA redeployed staff to support the hurricane response. Region IX officials added that as the region requested more staff, it sometimes received staff who had not been in the field for a while and lacked up-to-date knowledge to handle the mission needs, or staff who were coming off of long-term deployments from the hurricanes and were exhausted.

FEMA took several actions to address this shortfall in personnel by calling upon non-FEMA employees to deploy to the 2017 disaster areas. Specifically FEMA deployed more staff from other federal agencies through the use of the Surge Capacity Force, moved available staff between disaster zones as needed, hired locally, and used contractor personnel. For example, FEMA expanded the Surge Capacity Force for the first time in its history. Leveraging the Surge Capacity Force, the agency rapidly mobilized, trained, and equipped personnel from 34 non-DHS federal agencies to perform a variety of missions in support of the 2017 disasters, according to FEMA officials. FEMA officials in Florida said that the Surge Capacity Force was invaluable. One of our prior workforce recommendations was for FEMA to develop recruitment plans to address staff shortages in two new workforce elements, including the Surge Capacity Force.[70] FEMA concurred with this recommendation and has fully implemented it. However, FEMA officials said that they observed challenges with shifting personnel as conditions changed and said that the Surge Capacity Force still did not provide enough people, primarily due to the unanticipated staff demand created by the concurrent response to four major disasters.

---

[70] GAO-15-437. FEMA has implemented activities to address all recommendations in this report, including those related to the Surge Capacity Force.

Providing Training

FEMA recruited a large number of employees to meet the unprecedented demand for staffing resources in response to the 2017 disasters. According to FEMA officials, the large influx of new employees added to challenges conducting timely, program-specific training. More than half of FEMA personnel were serving in a capacity in which they did not hold the title of "Qualified"—during the peak of deployments to the 2017 disasters, according to our analysis of data from FEMA's Deployment Tracking System.[71] According to FEMA officials, an individual's qualification in the tracking system does not necessarily correlate to incomplete tasks and training. Officials added that there is no feasible way to provide the training required to fulfill all requirements for qualification because although some courses could be offered in the field, many staff positions require more than one training, not all of which can be delivered in a field environment.

According to the FCO for Hurricane Harvey, limited funding dedicated to training also hampered the agency's ability to meet basic training needs. This resulted in FEMA delivering aspects of the training even as the Surge Capacity Force actively responded to the disasters. FEMA officials stated that in the absence of required training, they developed just-in-time training, as well as hiring contractors to provide training locally. According to a FEMA official in Florida, FEMA should ideally provide training before a disaster so that the agency knows who is volunteering and can tailor training for tasks that fit their skills.

---

[71] According to FEMA officials, "Qualified" or "FQS qualification" is an official designation that refers to the fulfillment of criteria established by the FEMA Qualification System (FQS). The term is applied to FEMA personnel who, following an evaluation and validation of cadre-specific tasks and training requirements, are capable of independently executing their incident-based roles. The proportion of staff serving in a capacity in which they did not hold the title of "qualified" does not include non-FEMA disaster workforce components, such as the Surge Capacity Force and contractors. The qualification of these non-FEMA disaster workforce personnel is not tracked in FQS. For example, according to FEMA officials, Surge Capacity Force personnel are expected to self-report their skills when they arrive at their assigned location and receive training for roles that fit their self-assessment, prior to being deployed to the field to perform that role. If they are assigned a new role once they get to their duty station, they receive training for their new role. FEMA contractors as well as USACE members and local hires also receive program specific training before they begin working.

Retaining Staff

FEMA officials reported multiple challenges in retaining staff, including IMAT staff who play a key role in supporting the response to major disasters. According to FEMA officials, low pay and difficulty maintaining work-life balance have been identified as contributing to retention challenges on national Type I and regional Type II IMATs. According to the IMAT chief for Hurricane Harvey, attrition was high among IMAT staff because of the demanding nature of the job and low pay. FEMA officials in Florida also said that they faced challenges in hiring and retaining specialists such as engineers. According to these officials, the hiring process is lengthy and the pay is not competitive. Challenges revising and delivering training in a timely fashion may have also contributed to high turnover among reservists and other staff, according to the FCO in Texas. This is consistent with issues we had found in 2016 when we reported that the IMAT program had experienced high attrition across national and regional IMATs—since its implementation in fiscal year 2013—and that FEMA had not developed a strategy to address this challenge.[72] We recommended that FEMA develop a plan and workforce strategy for retention of IMAT staff. However, FEMA had not completed actions to respond to this recommendation as of July 2018. Nonetheless, FEMA officials said that the agency is examining its workforce challenges and has taken a number of actions to address staff retention. For example, officials told us they are working to expedite hiring decisions for IMAT personnel and increase the readiness of IMAT teams. According to FEMA officials, the agency has also made changes to ensure the composition of IMATs better reflect response priorities and foster greater team cohesion. Finally, it has established working groups to examine critical long-standing issues, such as training and equipment, that impact the team's ability to deploy and perform, in an effort to improve morale and retain personnel. We provide more details on the disaster workforce and related challenges in appendix V.

---

[72] GAO-16-87.

Use of Alternative Procedures and New Approaches to Service Delivery Offer Opportunities to Enhance Recovery Efforts but Also Present Challenges for the Public Assistance Program

Use of the alternative procedures and redesigned PA delivery model offers FEMA opportunities to help address some challenges experienced in past disasters and enhance overall recovery efforts. However, the relative lack of experience in administering disaster assistance using these approaches coupled with the magnitude of expertise and resource shortages presented by the response to multiple large-scale disasters also presents challenges.

Alternative Procedures

Local officials told us that, in consultation with FEMA, they are considering the extent to which they will use the Public Assistance Program alternative procedures to support their infrastructure recovery goals. As shown in Figure 14, the alternative procedures offer a modified approach to estimating and funding debris removal and permanent work under the Public Assistance programs in ways that can offer substantial benefits to disaster-affected communities.

According to FEMA officials, in the U.S. Virgin Islands and in the states, participation in the pilot program for permanent work alternative procedures will occur for select projects, by applicant request. These projects are to be administered using existing guidance that FEMA previously developed for its pilot program to implement the permanent work alternative procedures.[73] In contrast, for permanent work projects in Puerto Rico, alternative procedures will be used for all large permanent work projects.[74] On April 11, 2018, FEMA issued the *Public Assistance*

---

[73] FEMA updated this guide in 2016 and is applicable to all disasters declared on or after March 29, 2016. This guide is available at www.fema.gov/medialibrary/assets/documents/115868.

[74] Due to the extraordinary level of infrastructure damage caused by Hurricane Maria, as well as the financial status of Puerto Rico, officials chose to use the alternative procedures for all large project funding for Public Assistance categories C through G pursuant to section 428 of the Stafford Act. Puerto Rico; Amendment No. 5 to Notice of a Major Disaster

Alternative Procedures (Section 428) Guide for Permanent Work to guide recovery from Hurricane Maria in Puerto Rico after the Puerto Rico governor requested the use of alternative procedures for all large permanent work projects due to the unique circumstances in the territory (i.e., the magnitude of impacts and Puerto Rico's fiscal circumstances).

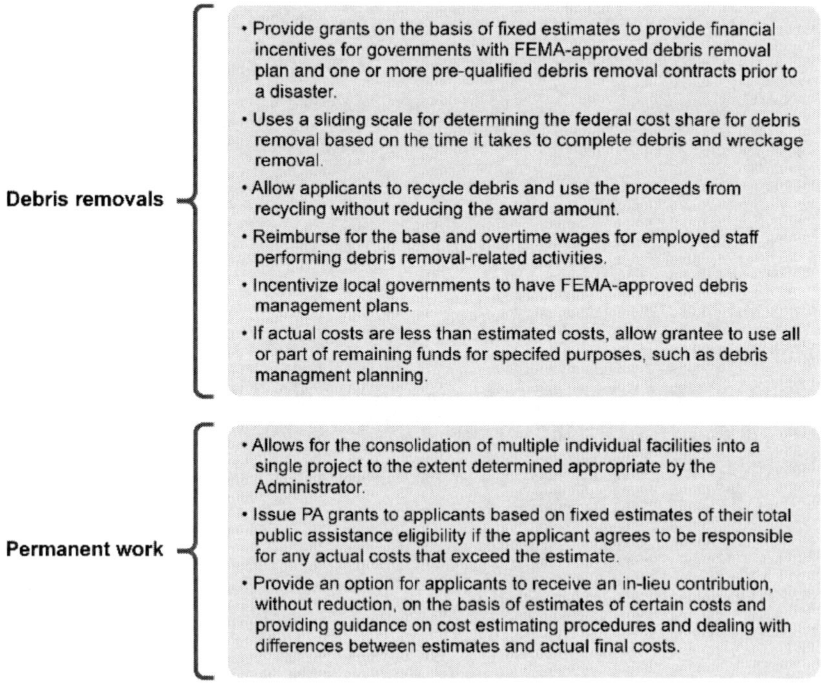

Source: GAO analysis of FEMA information. | GAO-18-472.

Note: See Pub. L. No. 113-2, div. B, § 1102(2), 127 Stat. 39, amending Pub. L. No. 93-288, tit. IV, § 428 (codified at 42 U.S.C. § 5189f).Following both Hurricane Katrina and Sandy, Congress authorized similar pilot programs to provide similar flexibilities in the Public Assistance Program, in an effort to reduce challenges and costs associated with the program, and incentivize states to speed the recovery process. FEMA operated that pilot program from June 2007 through December 2008 when authority for the pilot program expired.

Figure 14. Summary of Alternative Procedures for Federal Emergency Management Agency's (FEMA) Public Assistance Program, Authorized in 2013.

Declaration, 82 Fed. Reg. 53,514 (Nov. 16, 2017). For fiscal year 2018, the large project threshold is $125,500.

They added that this approach will provide the flexibility Puerto Rico requires to achieve its post-disaster recovery goals while limiting risk to the federal government.[75] According to FEMA and Puerto Rico officials, use of the alternative procedures is appropriate to support the large-scale rebuilding effort there. However, it is unclear whether such flexibilities will eliminate other challenges associated with the PA program, such as reducing delays from challenges to eligibility determinations and supporting a timely recovery. According to FEMA officials, although the front end of the PA alternative procedures pilot program may take longer than the standard PA procedures, once project formulation (including any identified hazard mitigation measures) and cost agreements are made, the entire sum of the agreed-upon funding level is obligated and made available to the recipient at a quicker rate due to the entire amount of the project being obligated from the outset. Additionally, the ability of an applicant to share funds between consolidated projects under alternative procedures reduces delays that occur when managing multiple complex projects coming in at higher or lower than anticipated costs, thus making for a smoother and more efficient recovery, according to FEMA officials. FEMA continues to develop a robust data collection and evaluation plan to perform substantive analysis of the pilot program and to better position itself to provide data on increased or decreased recovery timeframes, according to agency officials.

As of April 1, 2018, FEMA had approved 52 alternative procedures subawards in 30 states. Officials stated that the 52 awards likely represent large numbers of projects under the normal PA procedures, and therefore do not accurately reflect the frequency of FEMA and the states' use of the procedures. However, none of these projects were for recovery from a

---

[75] According to FEMA, any permanent work started before 12:00 am on September 17, 2017 will be attributed to Disaster Declaration 4336 in Puerto Rico, using the traditional PA procedures, and any work started after that time, regardless of whether the work was required by Hurricane Irma or Hurricane Maria, will be attributed to Disaster Declaration 4339 in Puerto Rico and will follow the procedures established in the disaster-specific guide. This guide describes the scope and limitations of the alternative procedures, the changes to the aspects of the PA Program to which these procedures apply, and identifies responsibilities for certain activities, as well as timelines for key actions and decisions. Where appropriate, FEMA may develop additional guidance and tools for implementation.

disaster the scale of the 2017 hurricanes and wildfires. Nonetheless, according to FEMA officials, the projects have provided valuable lessons learned which the agency has incorporated into its alternative procedures program guidance for Puerto Rico. We provide more details on the implementation of the PA program, including use of the alternative procedures, in appendix IX.

New Delivery Model

In recent years, FEMA has taken steps to redesign the PA program to address past challenges and make the program easier for FEMA and grantee officials to manage. As part of this effort, FEMA redesigned processes for developing, reviewing, and approving grant applications. The redesign also involves hiring for new PA staff positions, a standardized grant processing approach, and a new information system to better maintain and share grant documentation.

FEMA's original intention was to implement the new PA delivery model for all future disasters beginning in January 2018. However, in September 2017, FEMA expedited full implementation of the new model shortly after Hurricane Harvey made landfall. In a November 2017 report, we reviewed early implementation of the new delivery model in disasters that occurred prior to the four large-scale disasters of 2017 and found that FEMA needed to do more to assess the workforce needed to fully implement the model, such as the number of staff needed to fill certain new positions, or achieve staffing goals for supporting hazard mitigation on PA projects.[76]

We also found that FEMA had developed a new information-sharing system (now known as PA Grants Manager and Grants Portal) to address past information-sharing challenges, such as difficulties in sharing grant documentation among FEMA, state, and local officials and tracking the status of PA projects, but had not fully addressed two of four key information technology (IT) management controls—requirements development and systems testing and integration—that are necessary to

---

[76] GAO-18-30.

ensure systems work effectively and meet user needs. In our report, we recommended that FEMA assess the workforce needed for the new delivery model and improve the key IT management controls for Grants Portal. The agency concurred with our recommendations and according to officials, is taking steps to implement them. Our early observations about the workforce challenges experienced using the new delivery model in Texas, Florida, and California are similar to the challenges we cited in our prior work.[77] Although we saw some of the intended benefits of the redesign, officials in all eight municipalities we met with also cited one or more challenges with training and customer service, the new information system, or the incorporation of hazard mitigation into PA Projects, that have not yet been fully addressed.

Training and Customer Service

We interviewed local government officials from eight Texas, Florida, and California municipalities about their experiences using the new delivery model and officials from four of the municipalities said that they had positive experiences with the customer service provided by PA staff. For example, officials with a Texas municipality said that they have good communication with FEMA PA personnel whom they said are responsive to questions about Grants Portal. Officials from a Florida county said that under the old model they were forced to submit documents multiple times as a result of FEMA staff losing them, whereas with the new model, they can submit documents just once by uploading them directly into Grants Portal. However, local officials also noted challenges with the new delivery model. Specifically, officials with one California county said that they experienced staff turnover, and after about a month without any PA Program Delivery Manager, FEMA assigned a new official who was

---

[77] In our previous report, we found that officials who were testing the new delivery model for PA delivery experienced a variety of challenges. FEMA managers and PA applicants at the state and local level cited insufficient staff levels and problems with poorly trained staff, which affected customer service and the timely processing of PA grant applications from the states. Furthermore, we found deficiencies in how FEMA developed its new information system, which affected how well the system's capabilities would meet user needs and insufficiently addressed integration of the system with other internal and external information systems. See GAO-18-30.

untrained and inexperienced in the new process, and therefore unable to provide guidance on using the new IT system or answer questions about the process. Additionally, local officials in a Texas jurisdiction said that they received better customer service under the old model because the project manager writing the project worksheet could provide the applicant some information immediately to help gauge eligibility, and inform local fiscal decisions. Texas officials added that with the new process, all of the information is passed to centralized staff to write up the project while the Program Delivery Manager only collects project information, leaving the applicant unaware of next steps or reimbursement eligibility.

Speaking about their experiences using the new delivery model, officials in four municipalities we interviewed noted that they had little or no prior knowledge of these changes before using the new process for recovery from their 2017 disasters. Further, the officials stated that FEMA did not provide sufficient information about the new process. Texas, Florida, and California county officials we interviewed also expressed frustration with the lack of consistency in the implementation guidance FEMA provides and with project eligibility determination. According to officials from three of the eight jurisdictions, FEMA's process and decision-making is not clear and this is further complicated by the inability of FEMA staff to fully articulate the new process, due in part to limited training. FEMA officials acknowledged the challenge of training a large number of new employees on the new delivery model and stated that the agency has taken a number of steps to address these challenges and provide training to applicants and FEMA employees. For example, as part of its initial rollout of the new model, FEMA participated in "listening sessions" with state and local stakeholders. FEMA has also established and advertised a mechanism (the Change Control Tool) by which users of Grants Manager and Grants Portal can suggest a process improvement. Submissions to the tool are evaluated and prioritized on a monthly basis, according to FEMA officials. Additionally, officials said that FEMA is working to improve and expand its capacity to deliver training to applicants and FEMA employees. For example, in addition to hiring additional trainers, the agency is expanding the length and content of program training for field personnel, according to FEMA officials.

New Information System

Local government officials from two of the eight Texas, Florida, and California municipalities we interviewed about their experiences using Grants Portal commended the new PA information system for its transparency, however, officials from four jurisdictions expressed concerns about delays, not having sufficient guidance on how to use the new system, and how the system interfaces with their own state systems. Also, local officials from Texas, Florida and California expressed concerns about experiencing delays with FEMA's processing of their projects, and subsequent obligation of the necessary funds. Further, officials from three municipalities said it is unclear what happens once applicants submit their project applications to Grants Portal. For example, county officials in Florida said that under the old PA model, a FEMA project manager was available to walk applicants through the process of developing and submitting a project application; however, with the new model, applicants are required to develop and submit project applications without any assistance. Texas, Florida, and California officials said that after submitting their applications, they typically receive little information from FEMA about the processing of their application. According to FEMA officials, FEMA operates a hotline that FEMA staff and PA Program applicants can call to obtain answers to questions regarding use of the Grants Portal. FEMA is also working on further improving the functionalities of Grants Manager and Grants Portal to meet the needs of all of its users and to interface with the various federal and non-federal systems that track and manage PA grants. Additionally, expanding training will help improve both applicants' understanding of the process and FEMA employees' ability to work with applicants, according to FEMA officials.

Although it is too early to determine the effects of challenges with customer service and use of the information system under the new delivery model, local officials we met with have noted delays in processing PA applications. We provide more details on the experiences of each state using the new delivery model in appendix IX.

Incorporating Hazard Mitigation

The new delivery model under PA offers opportunities to better integrate hazard mitigation into recovery projects. Initial estimates indicate that the cost of damages from the 2017 hurricanes and wildfires will be far greater than the cost of damages for both hurricanes Katrina and Sandy. We have previously reported that using hazard mitigation and climate adaptation to enhance disaster resilience is critical to help address the federal fiscal exposure to disaster losses.[78] We also observed that although most funding for resilience activities is provided in the wake of a disaster, there are several aspects of incorporating resilience into post-disaster rebuilding efforts that may create barriers to its effectiveness. For example, in our prior work on Hurricane Sandy, we found that in some states, officials with primary responsibility for hazard mitigation noted that they wore other hats in the emergency operations center in the initial hours to days and were too focused on response functions to think about hazard mitigation.[79] FEMA designed the new PA delivery model to help state and local officials interact with FEMA experts to identify opportunities to incorporate hazard mitigation into PA projects early and throughout the process. Officials from 3 of the 6 municipalities we interviewed in Texas and Florida that are implementing the new delivery model confirmed that FEMA has provided mitigation experts to assist local officials with incorporating hazard mitigation on PA projects.[80] However, these officials also noted limitations in the extent to which mitigation experts were able to provide support on project development, echoing FEMA's broad workforce training challenges already discussed. According to officials with FEMA's PA Program Delivery Branch, FEMA is currently in the

---

[78] GAO-15-515.

[79] GAO-15-515.

[80] According to state officials in California, while disaster mitigation is a goal generally, due to the nature of the damages from these fires, the officials do not expect many opportunities for mitigation. California officials we spoke to did not state any challenges with hazard mitigation related to their PA recovery efforts.

process of conducting a comprehensive workforce review through which PA is assessing the size and composition of the deployable workforce. As part of this effort, FEMA is assessing the way it uses all available resources, including FEMA staff and technical assistance contractors to meet the demands of variable disaster cycles.

On site visits to Texas and Florida, we observed how instances of previous investments in disaster resilience reduced the damages from the storms, and may have lowered associated disaster costs. Both states and California have efforts underway that demonstrate their commitment to a resilient recovery. For example, Texas has a systematic approach to solicit hazard mitigation projects from local governments through Rebuild Texas—a commission set up by the Texas governor to marshal statewide resources and effort to rebuild public infrastructure damaged by Hurricane Harvey. Nevertheless, officials in Texas and Florida raised concerns about challenges to building disaster resilience during recovery, which were similar to the challenges we previously found reported by state and local officials affected by Hurricane Sandy.[81] These include the challenges with incorporating hazard mitigation that we previously discussed, and also challenges that go beyond PA projects, such as coordinating federal aid from multiple programs and local barriers to investment in resilience efforts. For example, Texas and Florida officials each described difficulties finding adequate funding to carry out key projects to make communities more resilient to flooding. We provide more details on how each state plans to incorporate disaster resilience, and their challenges, in appendix X.[82]

AGENCY COMMENTS

We provided a draft of this chapter to DHS and DOD for their review and comment. DHS provided a comment letter that is reprinted in appendix

[81] GAO-15-515.

[82] We plan to conduct future work on states' experiences incorporating resilience enhancements in response to and recovery from the 2017 disasters.

XIII and technical comments that we incorporated, as appropriate. DOD provided only technical comments that we incorporated, as appropriate.

In its comment letter, DHS acknowledged the challenges it faced conducting concurrent response operations for three major hurricanes—Harvey, Irma, and Maria—while preparing to take action on two more (hurricanes Jose and Nate) during the 2017 hurricane season. DHS also listed various actions it has taken since 2017 that are intended to expedite affected jurisdictions' recovery from the 2017 disasters, prepare for the 2018 hurricane season, and incorporate lessons learned to better prepare the nation for future disasters.

We will send copies of this chapter to the secretaries of Defense and Homeland Security, the FEMA Administrator, and appropriate congressional committees.

Chris P. Currie
Director
Homeland Security and Justice

List of Congressional Addressees

The Honorable Richard Shelby
Chairman

The Honorable Patrick Leahy
Vice Chairman
Committee on Appropriations
United States Senate

The Honorable Susan M. Collins
Chairman

The Honorable Robert P. Casey, Jr.
Ranking Member
Special Committee on Aging
United States Senate

The Honorable Mike Enzi
Chairman
Committee on the Budget
United States Senate

The Honorable Lisa Murkowski
Chairman

The Honorable Maria Cantwell
Ranking Member
Committee on Energy and Natural Resources
United States Senate

The Honorable Ron Johnson
Chairman

The Honorable Claire McCaskill
Ranking Member
Committee on Homeland Security and Governmental Affairs
United States Senate

The Honorable Rand Paul
Chairman

The Honorable Gary Peters
Ranking Member
Subcommittee on Federal Spending, Oversight
and Emergency Management
Committee on Homeland Security and Governmental Affairs
United States Senate

The Honorable Bill Nelson
United States Senate

The Honorable Marco Rubio
United States Senate

The Honorable Rodney Frelinghuysen
Chairman

The Honorable Nita Lowey
Ranking Member
Committee on Appropriations
House of Representatives

The Honorable Greg Walden
Chairman

The Honorable Joe Barton
Vice Chairman

The Honorable Frank Pallone, Jr
Ranking Member
Committee on Energy and Commerce
House of Representatives

The Honorable Jeb Hensarling
Chairman

The Honorable Maxine Waters
Ranking Member
Committee on Financial Services
House of Representatives

The Honorable Ann Wagner
Chairwoman

The Honorable Al Green
Ranking Member
Subcommittee on Oversight Investigations
Committee on Financial Services
House of Representatives

The Honorable Sean Duffy
Chairman

The Honorable Emanuel Cleaver
Ranking Member
Subcommittee on Housing and Insurance
Committee on Financial Services
House of Representatives

The Honorable Michael McCaul
Chairman

The Honorable Bennie Thompson
Ranking Member
Committee on Homeland Security
House of Representatives

The Honorable Dan Donovan
Chairman

The Honorable Donald Payne Jr.
Ranking Member
Subcommittee on Emergency Preparedness,
Response, and Communications
Committee on Homeland Security
House of Representatives

The Honorable Trey Gowdy
Chairman

The Honorable Elijah Cummings
Ranking Member
Committee on Oversight and Government Reform
House of Representatives

The Honorable Gary Palmer
Chairman
Subcommittee on Intergovernmental Affairs
Committee on Oversight and Government Reform
House of Representatives

The Honorable Greg Gianforte
Chairman
Subcommittee on the Interior, Energy, and Environment
Committee on Oversight and Government Reform
House of Representatives

The Honorable Bill Shuster
Chairman

The Honorable Peter DeFazio
Ranking Member
Committee on Transportation and Infrastructure
House of Representatives
Resident Commissioner Jenniffer González-Colón
House of Representatives

The Honorable John Culberson
House of Representatives

The Honorable Jeff Denham
House of Representatives

The Honorable Ron DeSantis
House of Representatives

The Honorable Alcee L. Hastings
House of Representatives

The Honorable Jared Huffman
House of Representatives

The Honorable Zoe Lofgren
House of Representatives

The Honorable Mike Thompson
House of Representatives

The Honorable Debbie Wasserman Schultz
House of Representatives

The Honorable Nydia Velázquez
House of Representatives

APPENDIX I: OBJECTIVES, SCOPE, AND METHODOLOGY

This chapter addresses the following: (1) our observations of the federal response coordination for hurricanes Harvey and Irma, in Texas and Florida, as well as for the California wildfires; (2) our observations of the federal response to hurricanes Irma and Maria in Puerto Rico and the U.S. Virgin Islands; and (3) existing and emerging disaster recovery challenges as well as opportunities highlighted by the 2017 hurricanes and wildfires.

To address all three objectives, we analyzed federal laws and FEMA policies, procedures, and guidance specific to emergency management. Specifically, we reviewed select sections of the Post-Katrina Act, including those associated with the establishment of (1) the *National Response Framework*, (2) the Federal Coordinating Officer (FCO) position—the lead federal official in charge of response, (3) Incident Management Assistance Teams (IMAT)—FEMA staff who rapidly deploy to an incident to provide leadership in the identification and provision of federal assistance and federal response capabilities, (4) the surge capacity force; and (5) the Sandy Recovery Improvement Act, particularly those sections associated with FEMA's public assistance program and debris removal responsibilities.[83] We also reviewed the *National Response Framework; the National Disaster Recovery Framework*; the 2017 *National Preparedness Report;* FEMA's 2014-2018 Strategic Plan, and the 2017 Hurricane Season FEMA After-Action Report. We included relevant information from our prior reports. Additionally, we obtained and analyzed key data from FEMA's financial management, workforce, emergency operations systems, and the federal procurement system. We focused on the highest period of disaster response activity for the federal government—August 2017 to January 2018, and we updated information provided as available in the subsequent months through June 2018. To ensure the reliability of the updated data, we interviewed officials at FEMA headquarters about their data quality control procedures, reviewed existing information about data systems—particularly data definitions and data validation, conducted electronic testing, and reviewed the data for obvious errors and omissions. We found the data to be sufficiently reliable for the purposes of this chapter.

---

[83] The *National Response Framework* is the part of the National Preparedness System established in Presidential Policy Directive 8 that is to be used to manage any type of disaster or emergency response, regardless of scale, scope, and complexity. Specifically, this framework covers actions to save lives, protect property and the environment, stabilize communities, and meet basic human needs following an incident. Response also includes the execution of emergency plans and actions to support short-term recovery. Department of Homeland Security, Federal Emergency Management Agency, *National Response Framework, Third Edition* (Washington, D.C.: June 2016). The surge capacity force is a cadre of non-FEMA federal employees who augment FEMA's disaster response and recovery efforts.

Moreover, in October and November 2017, we visited hurricane-damaged areas in the greater Houston area, throughout the southern part of Florida, in San Juan, Puerto Rico, and St. Croix, U.S. Virgin Islands. Further, we interviewed state emergency management officials or their designees in each disaster-affected state and territory as well as local government officials from eight municipalities in Texas, Florida, and California to gain their insights and perspectives on the federal response to the hurricanes and wildfires in their respective states and territories. We selected the cities and counties whose officials we interviewed based on their geographic proximity to the disaster affected sites we were already visiting, as well as their availability. The findings from these interviews cannot be generalized to all disaster-affected states; however, they provided insights to the respective states' response to and recovery from the disasters along with the federal government's role. We visited FEMA's joint field offices for: Hurricane Harvey, located in Austin Texas; Hurricane Irma, located in Orlando Florida; and hurricanes Irma and Maria, located in San Juan and St. Croix. We interviewed FEMA's on-site leadership in all of these locations, and conducted telephone interviews with FEMA's on-site leadership responsible for the response and recovery efforts in California. In addition, we conducted site visits to various FEMA branch offices and Disaster Response Centers (DRC) in Houston, Texas; Bonita Springs and Marathon, Florida; San Juan, Puerto Rico; and St. Croix, U.S. Virgin Islands.

To address our first objective of the federal response coordination for hurricanes Harvey and Irma, in Texas and Florida, as well as for the California wildfires, we:

- Conducted site visits to hurricane-impacted areas of Florida and Texas.
- Conducted interviews with: emergency management officials from California, Florida, and Texas; FEMA leadership from Regions IV, VI, and IX; officials from Houston, Texas, Harris County, Texas, Marathon, Florida, and Monroe County, Florida.

- Analyzed the implementation of the National Response Framework for hurricanes Harvey and Irma, and the California wildfires.

To address our second objective on the federal response to Hurricanes Irma and Maria in Puerto Rico and the U.S. Virgin Islands, we:

- Conducted site visits to hurricane impacted areas of Puerto Rico and U.S. Virgin Islands.
- Conducted interviews with officials from: the Puerto Rico Emergency Management Agency, the Puerto Rico Aqueduct and Sewer Authority, the Puerto Rico Electric Power Authority; the U.S. Virgin Islands Emergency Management Office; FEMA Region II; the Department of Defense (DOD) U.S. Northern Command (NORTHCOM); the U.S. Transportation Command (TRANSCOM); and the National Guard Bureau.
- Obtained information provided by DOD and FEMA on federal resources pre-positioned and provided in response to hurricanes Irma and Maria.

To address our third objective, existing and emerging disaster recovery challenges as well as opportunities highlighted by the 2017 hurricanes and wildfires; we:

- Conducted interviews with: emergency management officials from California, Florida, and Texas; FEMA leadership from Regions IV, VI, and IX; officials from Houston, Texas, Harris County, Texas, Marathon, Florida, and Monroe County, Florida.
- We obtained and analyzed data from FEMA's Deployment Tracking System (DTS)—FEMA's system for maintaining information on their workforce, including managing deployments of workforce elements to respond to declared disasters, and maintaining information on staff skills, qualifications, and training. To ensure that the data were sufficiently reliable for our purposes,

we interviewed officials at FEMA headquarters and regional offices about their data quality control procedures, reviewed existing information about data systems— particularly data dictionary and data validation, conducted electronic testing, and reviewed the data for obvious errors and omissions.

- We obtained and analyzed DHS and FEMA guidance related to federal response and recovery from major disasters. These include the 2017 *National Preparedness Report*; select DHS OIG Reports; our relevant prior reports; FEMA's *2014 – 2018 Strategic Plan; the National Disaster Recovery Framework*; FEMA's Alternative Procedures for Public Assistance Authorized in 2013 by the Sandy Recovery Improvement Act.

In addition, we undertook the following for the detailed information included in our appendices:

Federal Appropriations and FEMA Obligations for the 2017 Hurricanes and California Wildfires

- We obtained and analyzed data from FEMA's Integrated Financial Management Information System (IFMIS) on the amount of Disaster Relief Fund obligations and spend plans for the hurricanes and wildfires for fiscal years 2017 and 2018. IFMIS is FEMA's official accounting and financial management system that pulls all of FEMA's financial data from other FEMA, DHS, and government-wide systems (subsystems), and is the source of data for both internal and external financial reporting. The system records and tracks all financial transactions. To assess the reliability of these data, we reviewed the data and discussed data quality control procedures with FEMA officials. We determined that the data we used from these systems were sufficiently reliable for the purposes of this chapter.

We also obtained and analyzed disaster-related supplemental appropriations acts which provided additional direct funding for the 2017 hurricanes and wildfires included in our review.

Federal Response Coordination during the 2017 Hurricanes and Wildfires

- We obtained and analyzed FEMA's third edition of the National Response Framework (NRF). The NRF is a guide to how the nation responds to all types of disasters and emergencies. It establishes the federal response structure for disaster response that includes 14 core capabilities. We used the NRF to assess the FEMA-led federal response to the 2017 hurricanes and wildfires included in our review.
- We obtained and analyzed data from FEMA's web-based Emergency Operations Command (Web EOC)—the database used for submitting and tracking the disposition of resource requests from state and local officials to the federal government following a presidentially-declared disaster, among other things. To assess the reliability of these data, we reviewed the data for errors and discussed data quality control procedures with FEMA officials. We determined that the data we used from this system were sufficiently reliable for the purposes of this chapter.

Federal Contracting for the 2017 Disasters

- We obtained and analyzed data from FEMA, the U.S. Army Corps of Engineers, and the Federal Procurement Data System-Next Generation (FPDS-NG) through January 31, 2018, to determine federal contract obligations for each hurricane, the types of products and services procured, and rates of competition. We assessed the reliability of FPDS-NG data by reviewing existing information about the FPDS-NG system and the data it collects—specifically, the data dictionary and data validation rules—and performed electronic testing. We determined the FPDS-NG data were sufficiently reliable for the purposes of this chapter.

FEMA Disaster Workforce Capacity

- We obtained and analyzed data from DTS. To assess the reliability of these data, we reviewed the data and discussed data quality control procedures with FEMA officials. We determined that the data we used from these systems were sufficiently reliable for the purposes of this chapter.
- We obtained and analyzed FEMA's Human Capital Strategic Plan 2016-2020. The plan outlines FEMA's focus on five strategic goals specific to their workforce, and establishes goals for retention, training, recruitment, and capabilities gaps, among other things.

FEMA's Individual Assistance Program

- We obtained and analyzed FEMA's Individuals and Households Programs Unified Guidance –September 2016, which documents FEMA's policies and procedures for its Individual Assistance program—to include FEMA's housing program for individuals.
- Further, we obtained data from FEMA's Emergency Management Mission Integrated Environment (EMMIE)—on the number of individual Assistance (IA) applicants for the hurricanes and wildfires in our review; and we obtained and analyzed data from FEMA's IFMIS on the amounts obligated towards IA by FEMA as of February 28, 2018.
- We also obtained aggregated data from FEMA on the number of survivors who registered for disaster assistance as well as the

number who were approved for each type of housing and sheltering assistance, by disaster location as of June 22, 2018.

Fraud Risk Management in FEMA's Disaster Assistance Programs

- We obtained and analyzed information relevant to fraud risk management in FEMA's disaster-assistance programs. Specifically, we interviewed officials from FEMA's Fraud and Internal Investigations Division and obtained and analyzed relevant documentation including FEMA's Fraud Prevention and Investigation Directive. In addition, we reviewed information on FEMA's antifraud efforts reported in the DHS Fiscal Year 2017 Agency Financial Report. We also reviewed our relevant prior work and DHS Office of the Inspector General reports.

Payment Integrity Related to Disaster Relief Funding

- We analyzed information from our prior work on payment integrity and internal controls.

FEMA's Public Assistance Program

- We obtained PA obligation amounts from FEMA derived from EMMIE for California, Florida, Puerto Rico, Texas, and the U. S. Virgin Islands as of February 15, 2018.

Disaster Resilience and Hazard Mitigation

- We reviewed information from our prior work on the costs of natural disasters and the impact of hazard mitigation efforts. We obtained and analyzed information from FEMA and local officials on previous hazard mitigation efforts and future mitigation projects.

- We obtained and analyzed information from the draft National Mitigation Investment Strategy in 2018. The National Mitigation Investment Strategy provides a national approach to investments in mitigation activities and risk management across the United States.
- We obtained and analyzed information from FEMA's and HUD's guidance on hazard mitigation grant programs. This information details for what purpose and when hazard mitigation funding is available to states, tribes or territories.

Department of Defense's Support of Civil Authorities during the 2017 Hurricanes and Wildfires

- We conducted interviews with the following DOD components—Office of the Undersecretary for Defense Policy, Defense Support of Civilian Authorities (DSCA), NORTHCOM, TRANSCOM, Defense Logistics Agency, Army North, and Defense Coordinating Officers—who are the lead DOD representatives to FEMA during a major disaster declaration for FEMA Regions II, IV, VI, and IX. We also interviewed and collected data from the U.S. Army Corps of Engineers on their response efforts. We interviewed National Guard Bureau officials who provided personnel and logistical support for the 2017 disaster response efforts. During these interviews, we obtained information on their efforts to assist in the response to the 2017 hurricanes and wildfires. We also obtained summary data from these components on their response efforts. However, we did not verify the reliability of their data, although DOD officials noted that most of the data was also provided to external parties or made publicly available.

APPENDIX II: FEDERAL APPROPRIATIONS AND FEMA OBLIGATIONS FOR THE 2017 HURRICANES AND CALIFORNIA WILDFIRES

Overview

The Federal Emergency Management Agency's (FEMA) Disaster Relief Fund (DRF) is the primary source of federal disaster assistance for state and local governments. Congress appropriates no-year funding for the DRF, which allows FEMA to fund, direct, coordinate, and manage response and recovery efforts associated with domestic disasters and emergencies.

In addition to DHS, at least 29 other agencies carry out disaster assistance programs and activities. Some of these programs exist for the express purpose of supporting disaster response efforts and others are disaster-eligible—not specifically designated for disasters but able to be used for this purpose.

Federal Appropriations

Since the 2017 hurricanes and California wildfires, Congress and the President have provided the Department of Homeland Security (DHS) and 18 other federal agencies with at least $120 billion in supplemental appropriations for activities related to these disasters.[84] Table 2 summarizes these appropriations by agency.

[84] Supplemental Appropriations for Disaster Relief Requirements Act, 2017, Pub. L. No. 115-56, div. B, 131 Stat. 1129, 1136 (2017); Additional Supplemental Appropriations for Disaster Relief Requirements Act, 2017 Pub. L. No. 115-72, div. A, 131 Stat. 1224, 1224 (2017); Further Additional Supplemental Appropriations for Disaster Relief Requirements Act, 2018, Pub. L. No. 115-123, div. Supplemental Appropriations for Disaster Relief Requirements Act, 2018, Pub. L. No. 115-123, div. B, subdiv. 1, 132 Stat. 64, 65 (2018).

Table 2. Supplemental Appropriations for 2017 Hurricanes and California Wildfires

Agency	Purpose	Amount appropriated (in billions of dollars)[a]
Department of Homeland Security	For the Federal Emergency Management Agency Disaster Relief Fund, among other disaster-related purposes	50.7
Department of Housing and Urban Development	For restoration of infrastructure and housing and economic revitalization in impacted areas, among other disaster-related purposes	35.4
Department of Defense	For Army, Navy, Air Force and Marine Corps, expenses related to the 2017 hurricanes and Army Corps of Engineers construction of flood and storm damage reduction projects, among other disaster-related purposes	18.6[b]
Department of Agriculture	For expenses related to crops, trees, bushes, and vine losses; nutrition assistance; and wildland fire suppression operations, among other disaster-related purposes	5.5
Department of Education	For assisting in meeting the educational needs of individuals affected by a covered disaster or emergency, among other disaster-related purposes	2.7[c]
Small Business Administration	For direct loans to small businesses and associated expenses	2.1
Department of Transportation	For aviation, highway, and transit system expenses related to the consequences of the 2017 hurricanes, among other disaster-related purposes	1.8
Department of Health and Human Services	For Centers for Disease Control and Prevention expenses directly related to the consequences of the 2017 hurricanes, among other disaster-related purposes	1.1
Department of Commerce	For expenses related to flood mitigation, disaster relief, long-term recovery, and restoration of infrastructure in areas affected by the 2017 hurricanes, among other disaster-related purposes	1.0
10 other federal departments and agencies receiving less than $1 billion		1.2
Total		120.0[d]

Source: GAO analysis of disaster-related supplemental appropriations acts. I GAO-18-472.

Note: The supplemental appropriations provided up to $78.5 million for GAO and nine Inspectors General for oversight activities related to the expenditure of these funds.

[a]Some of the funds included in this table are no-year funds while others are available for specific periods of time. We do not include conditions, requirements, or authorities related to the use of funds.

[b]Does not include transfers of unobligated balances from prior fiscal years.

[c]Does not include indefinite appropriations authorized to forgive any outstanding balance owed to the Department of Education under the Historically Black College and University Hurricane Supplemental Loan program.

[d]Column does not sum to total due to rounding.

FEMA Disaster Relief Fund Obligations and Expenditures

FEMA reports monthly to Congress on its obligation and expenditure of DRF funds.[85] As of April 30, 2018, FEMA reports having obligated approximately $22.6 billion from the DRF for response and recovery efforts for the 2017 hurricanes and California wildfires, of which FEMA reports having expended a little over half—approximately $13.2 billion. FEMA projects obligating a total of about $30 billion by the end of fiscal year 2018.

Thus far, Hurricane Maria accounts for the highest amount of DRF obligations—$12.8 billion—followed by hurricanes Harvey and Irma. The 2017 California wildfires account for the smallest amount—approximately $1.4 billion obligated through April 30, 2018. See Figure 15.

FEMA tracks its DRF obligations in five categories—called programs:

1) Individual Assistance: The Individual Assistance Program provides financial assistance directly to disaster victims for the necessary expenses and serious needs that cannot be met through insurance or low-interest Small Business Administration loans, such as temporary housing assistance, counseling, unemployment compensation, or medical expenses.
2) Mission Assignment: Work orders FEMA issues that direct another federal agency to utilize its authorities and the resources granted to it under federal law in support of direct assistance to state, local, tribal, and territorial governments.
3) Public Assistance: The Public Assistance program provides financial assistance to state, tribal, territorial, and local governments for activities including debris removal; emergency

[85] An obligation is a definite commitment that creates a legal liability of the government for the payment of goods and services ordered or received. An expenditure is an amount paid by federal agencies, by cash or cash equivalent, during the fiscal year to liquidate government obligations. Whereas appropriations may, among other things, stipulate a federal agency's discretionary budget authority, obligations and expenditures are generally a better measure of the actual federal commitment (spending) of dollars on the provision of goods and services.

protective measures; and the repair, replacement, or restoration of disaster-damaged, publicly-owned facilities.

4) Administration: Costs for FEMA's delivery of disaster assistance including the salary and travel costs for the disaster workforce, rent and security expenses associated with field operation locations, and supplies and information technology for field operation staff, among other things.
5) Hazard Mitigation: The Hazard Mitigation Program provides funds to state, tribal, territorial, and local governments, among other entities, to assist communities in implementing long-term measures to help reduce the potential risk of future damages to facilities.

The proportion of funding obligated for each type of program varies as the disaster moves from more immediate response efforts to longer term response and recovery efforts. According to FEMA officials, soon after a disaster, Individual Assistance costs are typically the first to be obligated over a period of about 18 months. FEMA officials stated that mission assignment funds are also generally obligated during the response phase of a disaster, though the exact timing depends on how long it takes FEMA to receive and validate invoices from other federal agencies. Public assistance funds generally take longer to obligate—up to 10 years following a disaster—due to the complexity of funding for large public infrastructure projects, according to FEMA officials. Finally, FEMA officials stated that hazard mitigation funding is typically the last to be obligated, often many years after the disaster, as the response and recovery efforts abate.

The proportion of funding obligated for each type of program also varies due to other characteristics of the disaster. For example, mission assignment obligations comprised 60 percent of all actual and projected DRF obligations for the California wildfires and 4 percent of all such obligations for Hurricane Harvey. According to FEMA officials, neither the state of California nor the local counties had the capability or capacity to immediately take on the task of debris removal in response to the 2017 California wildfires, so FEMA funded this through a U.S. Army Corps of

Engineers mission assignment. In another example, the administration costs are projected to be highest for Hurricane Irma—31 percent—and lowest for the California wildfires—3 percent.

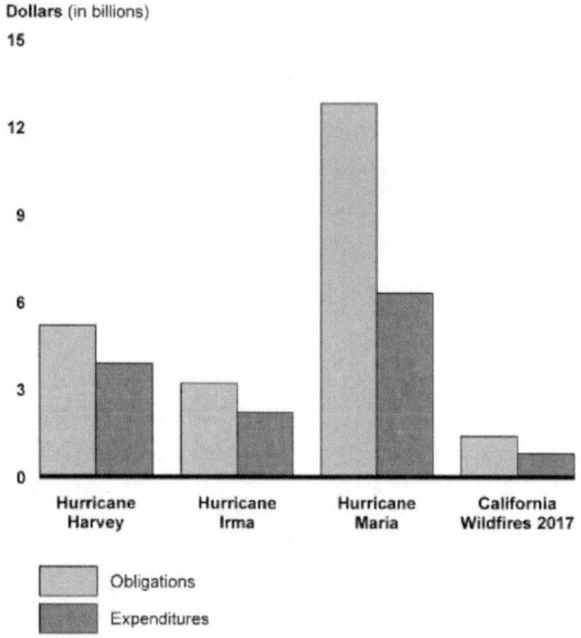

Source: GAO analysis of Federal Emergency Management Agency data. | GAO-18-472.

Notes: The data presented include obligations and expenditures incurred through April 30, 2018. Total obligations reported here include deobligations, meaning that they reflect cancellations or downward adjustments of an agency's previously incurred obligations. The Department of Homeland Security and other federal agencies have also spent non-Disaster Relief Fund funds responding to these disasters; these additional costs are not reflected here.

Figure 15. Disaster Relief Fund Obligations and Expenditures for Hurricanes Harvey, Irma, Maria, and California Wildfires, Through April 30, 2018.

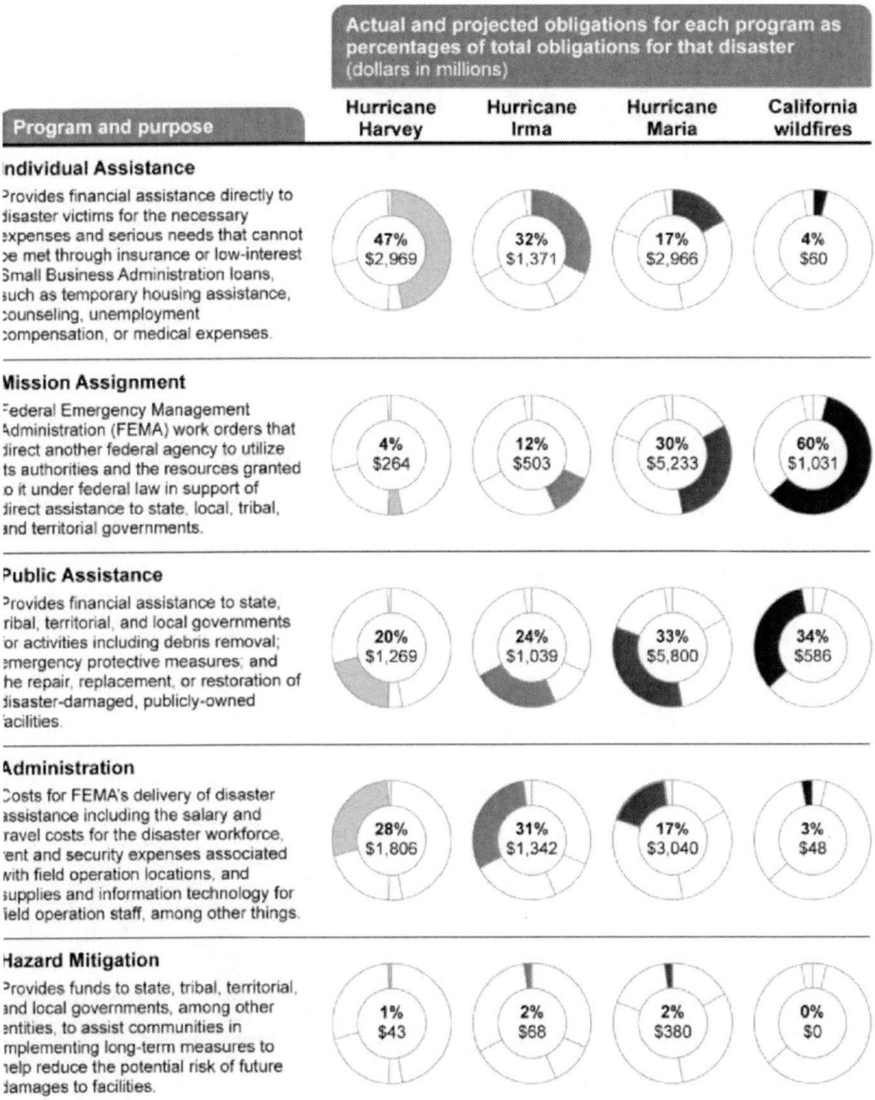

Source: GAO analysis of FEMA data. | GAO-18-472.

Note: These cost figures reflect obligations incurred as of April 30, 2018, and obligations projected through September 30, 2018. Mission assignment obligations include obligations for Urban Search and Rescue.

Figure 16. Actual and Projected Disaster Relief Fund Obligations For 2017 Hurricanes and California Wildfires, By Program and Disaster.

FEMA officials explained that the administrative costs for the 2017 California wildfires are much less because the majority of the wildfire costs were funded through mission assignments to the U.S. Army Corps of Engineers. These officials also stated that they expect the administrative costs for Hurricane Irma to decrease when program delivery levels off. See Figure 16 for the actual and projected DRF obligations by program.

APPENDIX III: FEDERAL RESPONSE COORDINATION DURING THE 2017 HURRICANES AND CALIFORNIA WILDFIRES

Overview

Major disasters such as the sequence of historically powerful hurricanes and damaging wildfires experienced in 2017, pose a challenge to national emergency preparedness.

The *National Response Framework* (NRF) states that the Secretary of Homeland Security is to ensure that preparedness actions are coordinated to prevent gaps in the federal government's efforts to respond to all hazards. Further, the NRF identifies Emergency Support Functions (ESF) that serve as the federal government's primary coordinating structure for building, sustaining, and delivering response capabilities. The ESFs define specific functional areas—such as communication, transportation, and energy—for the most frequently needed capabilities during an emergency to help coordinate the provision of assets and services by departments and agencies.

FEMA executes these support functions through "mission assignments," directing another federal agency to complete a specific task and citing funding, managerial controls, and guidance.

Federal Preparedness Activities Prior to Each Hurricane

Response activity for major disasters relies on coordination among federal, state, local and territorial governments, as well as on preparedness activities at all levels. The Federal Emergency Management Agency (FEMA), in coordination with the states and territories, took several actions in advance of the hurricanes in Texas, Florida, Puerto Rico and the U.S. Virgin Islands to help prepare for the disasters, as shown in Table 3.

Federal Response Activity for Each Disaster

There were 1,515 mission assignments—that is, an order from FEMA directing another federal agency to complete a specific task—for the 2017 hurricanes and California wildfires, and total obligations for these mission assignments was more than $7.8 billion, as of January 2018. Moreover, while all four incidents were declared major disasters—requiring Emergency Support Function (ESF) support that serve as the federal government's primary coordinating structure for building, sustaining, and delivering response capabilities, through mission assignments—the unique scale of the response activity in the territories is evident, as shown in Figure 17.

As of January 2018, the territorial response for hurricanes Irma and Maria alone resulted in more than 1,000 mission assignments and over $5.5 billion in obligations. Dozens of federal agencies provided support to FEMA and the states affected by the four disasters, but for all events, the Department of Defense (DOD) and the U.S. Army Corps of Engineers (USACE) were leading partners executing mission assignments from FEMA to support response activities. When requested, and approved by the Secretary of Defense, DOD provides Defense Support of Civil Authorities during domestic incidents and is therefore considered a support agency to all ESFs.

Table 3. Examples of Federal Disaster Preparedness Activities Undertaken Prior to the 2017 Hurricanes, By Location

Preparedness Activity	Hurricane Harvey in Texas as of August 25, 2017	Hurricane Irma in Florida as of September 9, 2017	Hurricane Irma in Puerto Rico and the U.S. Virgin Islands as of September 6, 2017	Hurricane Maria in Puerto Rico and the U.S. Virgin Islands as of September 19, 2017
Urban Search and Rescue Personnel Deployed	6 task forces	1,303	1,430	276
Incident Management Assistance Teams Deployed[a]	1 national 3 regional	0 national 1 regional	2 regional	1 national 2 regional
Federal Workforce Deployed	784	16,639	2,257	2,763
Disaster Medical Assistance Teams Deployed	0	13	1	0
Number of Generators Delivered	35	0	24	4
Commodities Delivered	306,966 Meals and 96,978 liters of water delivered	4.8million meals and 9.9M liters of water delivered	1 million meals, and 2 million liters of water delivered	1,617,241meals, and698,570 liters of water delivered
Open Shelters	7 shelters open in a with population of 91	249 shelters open with a population of 48,739	26 shelters open with a population of 388	8 shelters open with a population of 306
Number of Emergency Support Functions Activated[b]	13 of 14	13 of 14	12 of 14	13 of 14

Source: GAO Analysis of Federal Emergency Management Agency's Senior Leadership Briefing and Recovery Snapshots for the 2017 disasters. I GAO-18-472.

Notes:

[a]Incident Management Assistance Teams are FEMA staff in areas such as operations, logistics, planning, and finance and administration that rapidly deploy to an incident or incident-threatened venue to provide leadership in the identification and provision of federal assistance, and coordinate and integrate inter-jurisdictional response in support of an affected state, tribe, or territory.

[b]Emergency Support Functions serve as the federal government's primary coordinating structure for building, sustaining, and delivering response capabilities.

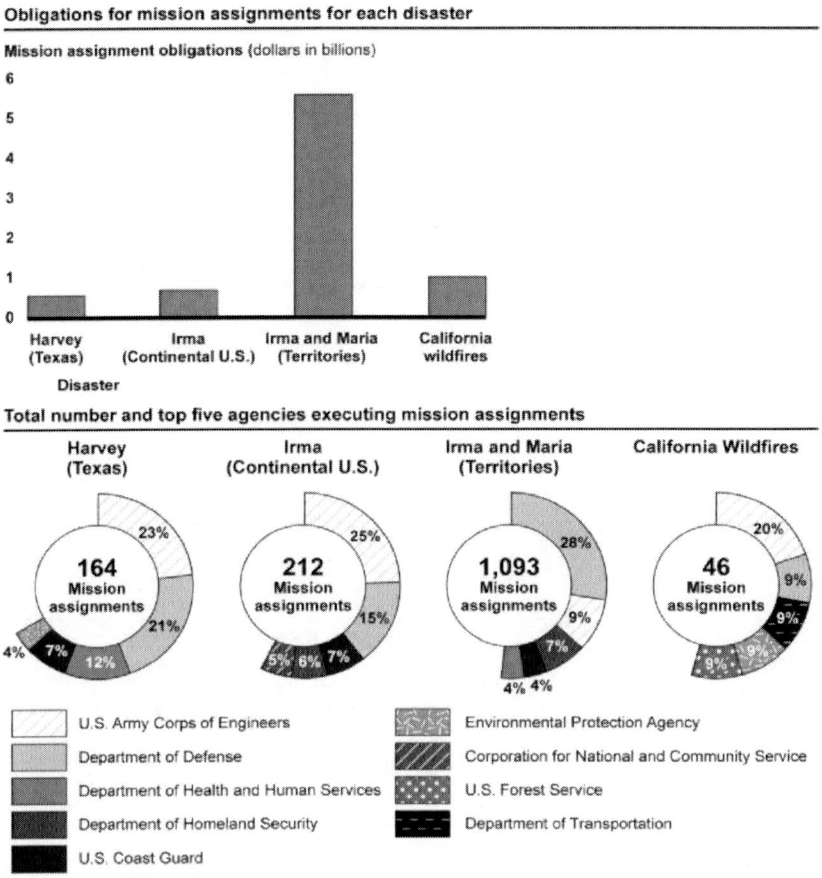

Source: GAO analysis of data from the Federal Emergency Management Agency's Web-based Emergency Operations Center (WebEOC). | GAO-18-472.

Note: Mission assignment—that is, an order from the Federal Emergency Management Agency (FEMA) directing another federal agency to complete a specific task—totals and percentages for each agency reflect data entered into FEMA's Web-based Emergency Operations Center (WebEOC) system of approved resource requests by response officials. Data presented reflect totals and percentages where data are available. This does not include 1,285 request records (out of 3,339 total) for which data are not available to determine whether FEMA sourced the request through a mission assignment or other source, and 74 records for which data are not available to determine which agency FEMA assigned out of the 1,515 mission assignments reviewed.

Figure 17. Federal Government Response Activity and Obligations through Mission Assignments for Each 2017 Disaster, as of January 2018.

Table 4. Number of Mission Assignments Received by Emergency Support Function (ESF) Agencies in Response to the 2017 Disasters and Selected Examples of Support, as of January 29, 2018

ESF #1: Transportation Coordinator: Department of Transportation		
Harvey	3	In response to Hurricane Harvey in Texas, the Department of Transportation and the Federal Transit Administration received a mission assignment from the Federal Emergency Management Agency (FEMA) deploying personnel to assist with assessing the damages to public transit systems and associated costs.
Irma in the Mainland	4	
Irma and Maria in the territories	51	
California Wildfires	4	
ESF #2: Communications Coordinator: DHS /National Communications System		
Harvey	2	In response to Hurricane Irma in Florida, the General Services Administration received a mission assignment from FEMA deploying personnel to assist with disaster response operations at various locations.
Irma in the Mainland	2	
Irma and Maria in the territories	14	
California Wildfires	2	
ESF #3: Public Works and Engineering Coordinator: U.S. Army Corps of Engineers		
Harvey	35	In response to the hurricanes in the U.S. Virgin Islands, the U.S. Army Corps of Engineers received a mission assignment from FEMA to provide temporary roofing for the territory in support of response operations. This included deploying the temporary roofing team and implementing contracting processes to provide temporary support.
Irma in the Mainland	44	
Irma and Maria in the territories	96	
California Wildfires	8	
ESF #4: Firefighting Coordinator: U.S. Forest Service		
Harvey	4	In response to the California wildfires, the U.S. Forest Service and the Department of Interior received a mission assignment from FEMA to provide fire assistance and suppression implementation planning.
Irma in the Mainland	3	
Irma and Maria in the territories	31	
California Wildfires	5	
ESF #5: Information and Planning Coordinator: FEMA		
Harvey	4	In response to the flooding related to Hurricane Harvey, the U.S. Geological Survey received a mission assignment from FEMA to provide advance support, real-time field measurements, and daily reporting of water heights for counties along the Gulf of Mexico coast.
Irma in the Mainland	1	
Irma and Maria in the territories	74	
California Wildfires	1	

ESF #6: Mass Care, Emergency Assistance, Housing, and Human Services Coordinator: FEMA		
Harvey	4	In response to Hurricane Irma in Florida, the Department of Housing and Urban Development received a mission assignment from FEMA to support multiple programs providing shelter to disaster survivors. For example, assessing and coordinating assistance to elderly populations in the Transitional Shelter Assistance Program, coordinating with public housing authorities on timelines for repairs to damaged units; and efforts to ensure Fair Housing Act compliance.
Irma in the Mainland	5	
Irma and Maria in the territories	20	
California Wildfires	0	
ESF #7: Logistics Management Coordinators: General Services Administration and FEMA		
Harvey	5	In response to hurricanes Irma and Maria in Puerto Rico, the Defense Logistics Agency received a mission assignment from FEMA to provide 82 million commercial "meals ready to eat" in support of response operations.
Irma in the Mainland	5	
Irma and Maria in the territories	122	
California Wildfires	3	
ESF #8: Public Health and Medical Services Coordinator: Department of Health and Human Services		
Harvey	18	In response to damages from Hurricane Maria on the island of St. Croix, in the U.S. Virgin Islands, the DOD received a mission assignment from FEMA to provide deployable temporary medical facilities to the island.
Irma in the Mainland	7	
Irma and Maria in the territories	70	
California Wildfires	2	
ESF #9: Search and Rescue Coordinator: FEMA		
Harvey	13	In response to the historic flooding caused by Hurricane Harvey in Texas, multiple agencies—including U.S. Customs and Border Protection, U.S. Coast Guard, and the Department of Fish and Wildlife—received mission assignments from FEMA to provide boating equipment to move up to 20,000 survivors.
Irma in the Mainland	5	
Irma and Maria in the territories	15	
California Wildfires	1	
ESF #10: Oil and Hazardous Materials Coordinator: Environmental Protection Agency		
Harvey	7	In response to Hurricane Irma in Florida, the Environmental Protection Agency received a mission assignment from FEMA to support the assessment and response operations to actual or threatened hazardous substances and oil releases to remove the threat of danger or contamination to the public.
Irma in the Mainland	14	
Irma and Maria in the territories	27	
California Wildfires	4	
ESF #11: Agricultural and Natural Resources Coordinator: Department of Agriculture		
Harvey	2	In response to hurricanes Irma and Maria, the U.S. Department of Agriculture received a mission assignment from FEMA to provide personnel with technical expertise in responding to animal and agricultural health issues, agricultural emergency management, and nutrition assistance in support of response operations in Puerto Rico.
Irma in the Mainland	4	
Irma and Maria in the territories	15	
California Wildfires	3	

Table 4. (Continued)

ESF #12: Energy Coordinator: Department of Energy		
Harvey	1	The U.S. Department of Energy received a mission assignment from FEMA to provide subject matter experts in electrical distribution, transmission, generation, energy efficiency, renewable energy and related topics to advise the U.S. Army Corps of Engineers on the assessment, planning and reconstruction of the electrical grid in Puerto Rico following Hurricane Maria.
Irma in the Mainland	3	
Irma and Maria in the territories	32	
California Wildfires	3	
ESF #13: Public Safety and Security Coordinator: Department of Justice		
Harvey	7	The Federal Protective Services received a mission assignment from FEMA to provide security guard service at FEMA facilities in Southern California, following the 2017 wildfires.
Irma	3	
Maria	39	
Wildfires	4	
ESF #15: External Affairs Coordinator: DHS		
Harvey	0	DOD received a mission assignment from FEMA to provide support to distribute emergency messages in support of Hurricane Irma response operations in the U.S. Virgin Islands.
Irma	0	
Maria	49	
Wildfires	0	

Source: FEMA. I GAO-18-472.

Note: Mission assignment data for each ESF and disaster reflect data entered into FEMA's Web-based Emergency Operations Center (WebEOC) system of approved resource requests by response officials. This does not include 591 records for which data are not available to determine which ESF response officials identified for the mission assignment. ESF data for Maria includes obligations that stemmed from Hurricane Irma's impact on the U.S. Virgin Islands and Puerto Rico. ESF 14 is no longer in use as of 2011. When requested, and approved by the Secretary of Defense, DOD provides Defense Support of Civil Authorities during domestic incidents and is therefore considered a support agency to all ESFs. In some cases, DOD received a greater percentage of requests under the ESF than the lead coordinating agencies. According to FEMA Office of Response and Recovery officials, the 2017 disasters challenged FEMA in many ways, necessitating a larger role for DOD due to its specialized capabilities. Further, the response efforts followed the FEMA Region II Puerto Rico and U.S. Virgin Islands Hurricane Annex planning considerations, which identifies DOD for several mission responsibilities due to the unique considerations of the Islands' location. In addition, DOD was needed for missions deemed "uncommon" such as airfield assessments and opening.

In some cases, DOD received a greater percentage of requests under the ESF than the lead coordinating agencies. For example, in response to hurricanes Irma and Maria in the territories, FEMA assigned DOD 35 requests and USACE one request for ESF 8, for which the Department of Health and Human Services (HHS) is the lead coordinating agency; whereas FEMA assigned HHS 32 requests for ESF 8. According to FEMA Office of Response and Recovery officials, the 2017 disasters challenged FEMA in many ways, necessitating a larger role for DOD due to its specialized capabilities. Further, the response efforts followed the FEMA Region II Puerto Rico and U.S. Virgin Islands Hurricane Annex planning considerations, which identifies DOD for several mission responsibilities due to the unique considerations of the Islands' location. In addition, DOD was needed for missions deemed "uncommon" such as airfield assessments and opening. For each ESF— there is a lead federal agency designated as the coordinator. The ESF coordinators oversee the preparedness activities for a particular ESF and coordinate with its primary and support agencies. See Table 4 for selected examples of ESF delivered in support of all four 2017 disasters.

APPENDIX IV: FEDERAL CONTRACTING FOR THE 2017 HURRICANES

Overview

Historically, federal contracts comprise a large share of federal expenditures for hurricane response and recovery efforts. It can take years to fully account for federal contract obligations resulting from a hurricane. For example, federal agencies are still making contract obligations as part of the recovery efforts as far back as hurricanes Sandy and Katrina, which occurred in 2012 and 2005, respectively. According to early estimates, the 2017 hurricanes are among the most expensive hurricanes in terms of federal contract obligations since 2005, when agencies began tracking information by hurricane.

Advance Contracting

To facilitate a faster response, FEMA and USACE identify goods and services that are typically needed for disaster response and establish contracts for them—known as advance contracts—prior to the disasters. As of January 31, 2018, FEMA reported that it had obligated over $2.4 billion through advance contracts for products and services such as prefabricated buildings, food, and inspection services, in response to hurricanes Harvey, Irma, and Maria. As of the same date, USACE reported that it obligated about $555 million through its advance contracts for services such as temporary power, temporary roofing, and debris removal.

Competition

Across all three hurricanes, we found that as of January 31, 2018, the overall competition rate—the percentage of total obligations reported under competitive contracts—was 81 percent.

Total Contract Obligations for Hurricanes Harvey, Irma, and Maria

A number of federal departments procured goods and services in response to the three 2017 hurricanes. As of January 31, 2018, the Department of Homeland Security (DHS), which includes the Federal Emergency Management Agency (FEMA), and the Department of Defense (DOD), which include the U.S. Army Corps of Engineers (USACE), accounted for approximately 96 percent of total contract obligations[86]

[86] For the purposes of this appendix, contract obligations include obligations against what the General Services Administration's Federal Procurement Data System-Next Generation (FPDS-NG) categorizes as definitive vehicles (definitive contracts and purchase orders that have a defined scope of work that do not allow for individual orders under them), and against what FPDS-NG categorizes as indefinite delivery vehicles (orders under the Federal Supply Schedule, orders/calls under blanket purchase agreements, orders under basic

across 21 federal departments.[87] As of January 31, 2018, federal departments had obligated over $6.2 billion for contracts in support of the response and recovery efforts for hurricanes Harvey, Irma, and Maria. Figure 18 provides details on agencies' contract obligations in support of the three hurricanes.

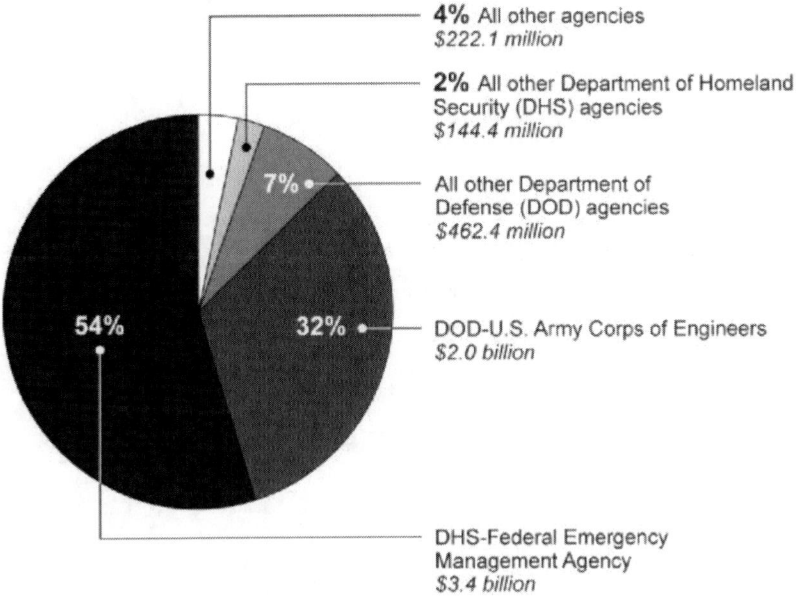

Source: GAO analysis of Federal Procurement Data System-Next Generation data. | GAO-18-472.

Figure 18. Contract Obligations in Support of Hurricanes Harvey, Irma, and Maria Response Efforts, by Agency, through January 31, 2018.

ordering agreements, orders under government-wide acquisition contracts, and orders under other indefinite delivery vehicles, such as indefinite delivery, indefinite quantity contracts).

[87] In addition to DOD and DHS, the following departments had contract obligations in support of hurricanes Harvey, Irma, and Maria response efforts: the departments of Agriculture, Commerce, Energy, Health and Human Services, Housing and Urban Development, Interior, Justice, State, Transportation, Treasury, and Veterans Affairs; the Agency for International Development; the Broadcasting Board of Governors; the Corporation for National and Community Service; the Environmental Protection Agency; the General Services Administration; the National Aeronautics and Space Administration; the Social Security Administration; and the Small Business Administration.

Observations on Contract Obligations by Hurricane

Each of the three 2017 hurricanes hit different geographic locations and caused varying degrees of destruction, from flooding and wind damage to massive power outages. As such, contract obligations varied by hurricane in terms of amount and whether they were for products or services, as shown in Table 5.

Table 5. Contracting Information by 2017 Hurricane through January 31, 2018

	Harvey	Irma	Maria
Total Contract Obligations (Dollars in millions)	1,364	1,026	3,826
Percent obligated by Federal Emergency Management Agency (%)	79	71	41
Percent obligated by U.S. Army Corps of Engineers (%)	4	3	50
Contract obligations for products (Dollars in millions)	649	213	1,259
Contract obligations for services (Dollars in millions)	715	813	2,567
Competition Rate (%)	82	7%	81

Source: GAO analysis of Federal Procurement Data System-Next Generation data. I GAO-18-472

Note: For the purposes of this appendix, competition rate is the percentage of total obligations associated with contracts awarded competitively. We calculated competition rates as the percentage of obligations on competitive contracts over all obligations on contracts annually.

Numbers may not add due to rounding. Examples of products procured through contracts include food, water, and shelter; while examples of services include power restoration and the repair or alteration of damaged buildings.

Competitive contracts included contracts and orders coded in the Federal Procurement Data System-Next Generation (FPDS-NG) as "full and open competition," "full and open after exclusion of sources," and "competed under simplified acquisition procedures," as well as orders coded as "subject to fair opportunity" and as "fair opportunity given," and "competitive set aside." Noncompetitive contracts included contracts and orders coded in FPDS-NG as "not competed," "not available for competition," and "not competed under simplified acquisition procedures," as well as orders coded as an exception to "subject to fair opportunity," including "urgency," "only one source," "minimum guarantee," "follow-on action following competitive initial action," "other statutory authority," and "sole source."

Approximately $3 billion of the $6.2 billion in contracts obligated for the three hurricanes as of January 31, 2018, was obligated through advance contracts. Table 6 provides additional details on FEMA and USACE obligations on advance contracts.

Table 6. Information on Advance Contracting by 2017 Hurricane, through January 31, 2018

	Harvey	Irma	Maria
Obligations on Advance Contracts (Dollars in millions)	948	566	1,493
Federal Emergency Management Agency (Dollars in millions)	940	536	975
U.S. Army Corps of Engineers (Dollars in millions)	7	30	518

Source: GAO analysis of Department of Homeland Security and Department of Defense data. I GAO-18-472.

Note: Numbers may not add due to rounding. Advance contracts are contracts identified and established prior to a disaster, for goods and services that are typically needed during a disaster response.

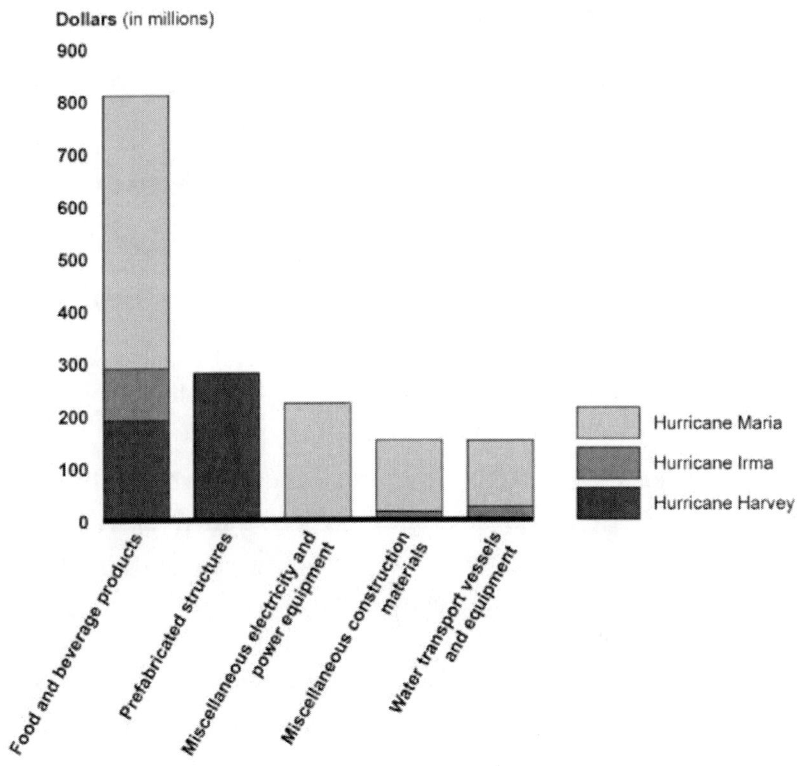

Source: GAO analysis of Federal Procurement Data System-Next Generation data. | GAO-18-472.

Figure 19. Top Five Product Groups in Terms of Contract Obligations through January 31, 2018, and Proportion of Obligations by Hurricane.

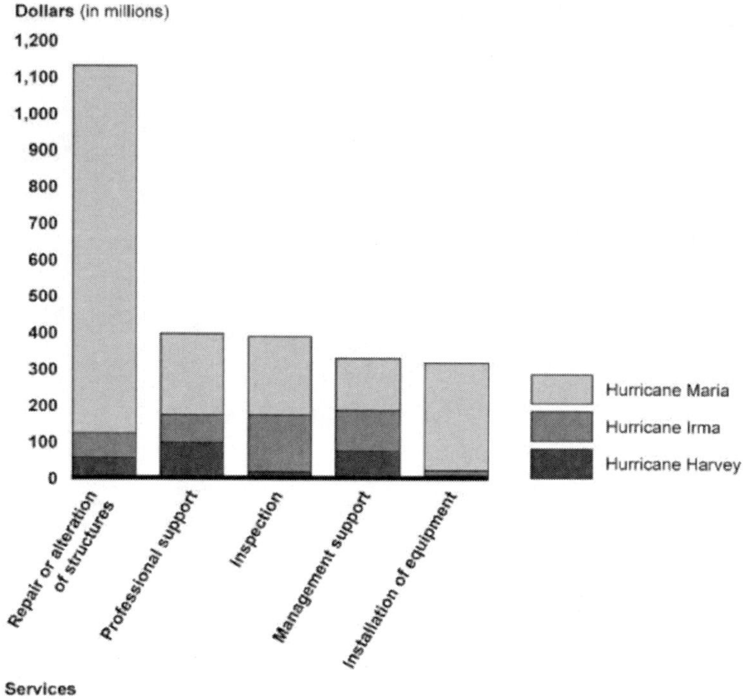

Source: GAO analysis of Federal Procurement Data System-Next Generation data. | GAO-18-472.

Figure 20. Top Five Service Groups in Terms of Contract Obligations through January 31, 2018, and Proportion of Obligations by Hurricane.

Federal agencies procured a variety of products and services through contracts in response to the hurricanes, obligating more than $2.1 billion for products and about $4.1 billion for services. Figure 19 identifies the top five product groups in terms of contract obligations, and the proportion of obligations for each hurricane. These contracts include life-sustaining products such as food, water, and power for survivors.

Figure 20 identifies the top five service groups in terms of contract obligations, and the proportion of obligations for each hurricane. For example, these contracts include inspection services, such as housing inspections, and professional support services, such as support for FEMA's housing and feeding missions in the affected areas.

Additional details on disaster contracting and on obligations for products and services procured for the three hurricanes can be found in our prior work on 2017 disaster contracting.

APPENDIX V: FEMA DISASTER WORKFORCE CAPACITY

Overview

Under the Stafford Act, FEMA has the authority to augment its permanent full-time staff with temporary personnel when needed. Additionally, during a disaster response, FEMA deploys nonFEMA employees from two workforce components—the Surge Capacity Force and the FEMA Corps. FEMA also hires locally and employs other personnel, such as contractors, to provide a variety of products and services, such as debris removal.

According to FEMA's 2014-2018 Strategic Plan, the agency's goal is to develop and manage its disaster workforce to respond to two concurrent catastrophic disasters. Although the federal disaster workforce FEMA can deploy has expanded in recent years, to over 24,000 as of August 20, 2017, the agency faced challenges training employees and maintaining staffing levels across four concurrent disasters.

In September 2017, in response to staffing shortfalls, FEMA expanded its Surge Capacity Force program to include not only volunteers from the Department of Homeland Security (DHS), but all federal employees. As of January 2018, the program had enrolled over 12,000 employees, compared to 4,033 in 2015.

FEMA Faced Challenges Responding to Sequential Disasters in Late 2017

The Federal Emergency Management Agency (FEMA) experienced challenges in recruiting and maintaining its workforce to support the hurricanes and wildfires response in 2017. Prior to landfall of Hurricane Harvey in August 2017, FEMA had already deployed staff to other long-term recovery operations. In addition, based on its internal workforce analyses, FEMA faced a staff shortage of 37 percent as of September 1, 2017. Figure 21 shows the deployment of the federal disaster personnel across the 2017 disasters.

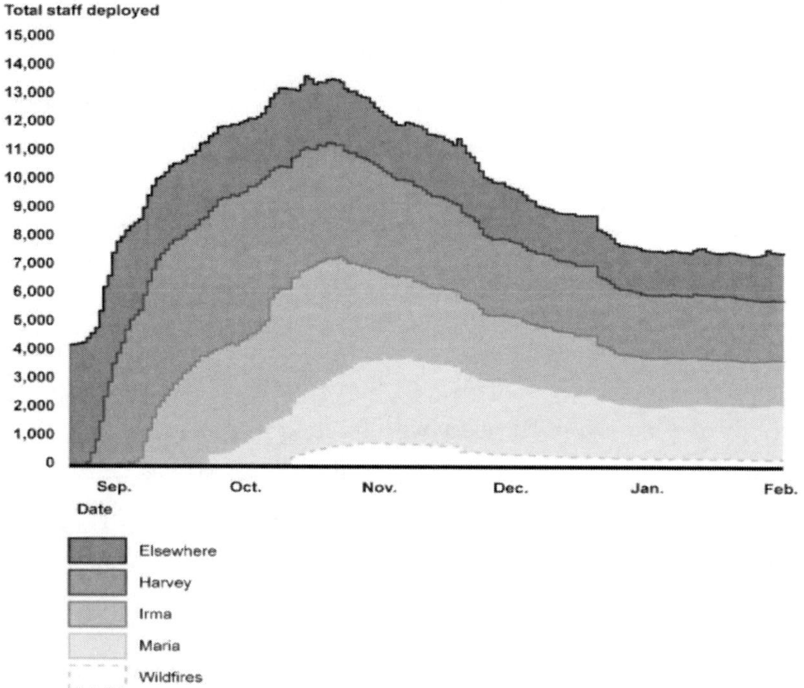

Source: GAO analysis of Department of Homeland Security data. | GAO-18-472.

Note: Data include Federal Emergency Management Agency (FEMA) workforce and non-FEMA employees that FEMA can deploy during a disaster response such as the Surge Capacity Force, FEMA Corps and contractors. Data do not represent local hires or employee types such as permanent part-time, temporary part-time, and temporary incident employees.

Figure 21. Federal Disaster Workforce Deployed In Response to the 2017 Disasters from September 2017 through February 2018.

Less than Half of FEMA's Deployed Workforce Held a Qualified Title during 2017 Disasters

FEMA faced challenges maintaining a "Qualified" workforce—a FEMA Qualification System designation that refers to personnel who, following an evaluation and validation of specific tasks and training requirements, are capable of independently executing their specific roles—across concurrent disasters. As shown in Figure 22, at the height of workforce deployments in mid-October 2017, 54 percent of staff were serving in a capacity in which they did not hold the title of "Qualified."

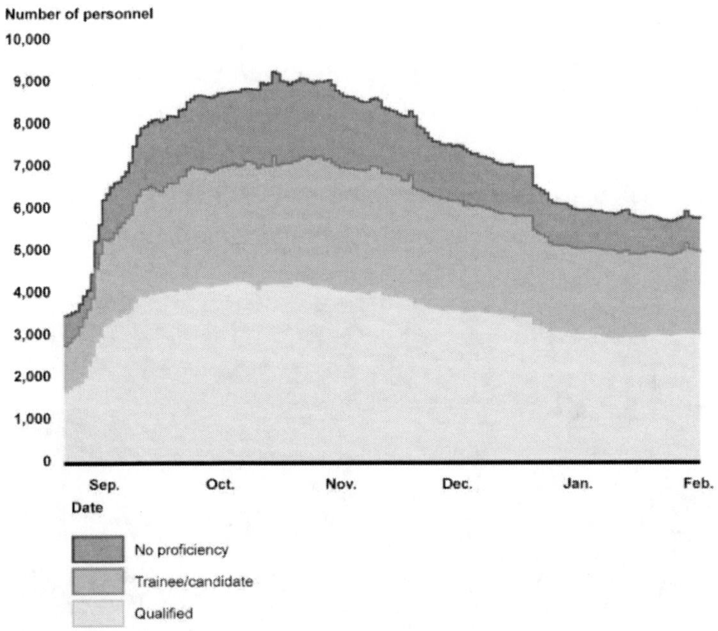

Source: GAO analysis of Department of Homeland Security data. | GAO-18-472.

Note: Data only represent FEMA employees and do not include local hires, FEMA Corps, Surge Capacity Force, contractors or employees from other federal agencies.

Figure 22. Federal Emergency Management Agency Qualification Levels for Deployed Staff during the 2017 Disaster Response.

Observations from Affected States and Territories

Challenges Reported by Officials in Responding to Sequence of 2017 Disasters

Local and FEMA officials from Texas and California—the two states that experienced disasters at the beginning and end of the 2017 disasters season—expressed concern over FEMA's workforce capacity in responding to concurrent catastrophic disasters. FEMA and local officials in Texas said that staff were initially deployed when Hurricane Harvey landed, but some staff were re-deployed once hurricanes Irma and Maria hit. Local officials in California described the difficulty in responding to wildfires with a shortage of staff, while FEMA officials noted the exhaustion of staff re-deployed to California after returning from deployments in the areas impacted by the hurricanes—often with only a 1- or 2-day break in between. As a result, many staff were not in a state to best serve mission needs according to officials. An employee serving at the start of the 2017 hurricane season is shown in Figure 23.

Source: FEMA. | GAO-18-472.

Figure 23. Federal Emergency Management Agency (FEMA) Employee in Texas.

Mixed Results in Matching Surge Capacity Force Skills to Job Assignments

The Surge Capacity Force comprises DHS and other federal employees who volunteer to deploy in the event of a disaster for a maximum of 45 days of service.[88]

- Available staff as of August 2017: 6,537
- Peak deployed during 2017 disasters: 3,102

Hurricane Harvey (Texas)

FEMA officials reported successes with utilizing the expertise of the Surge Capacity Force to match disaster needs by pre-collecting data on skill sets.

Hurricane Irma (Florida)

FEMA officials noted the value of the Surge Capacity Force, but cited challenges with understaffing and identifying tasks that best match their skill sets (e.g., a NASA engineer could be better used to help with planning rather than loading copy paper into printers). Figure 24 shows a volunteer in Florida.

Hurricane Maria and Hurricane Irma (Puerto Rico)

FEMA officials noted the success of the Surge Capacity Force, but volunteers cited concerns in matching their skill sets to disaster recovery tasks in Puerto Rico as well as disconnects between information given during training and the job requirements.

Hurricane Maria and Hurricane Irma (U.S. Virgin Islands)

FEMA officials in the U.S. Virgin Islands noted the positive attitude and integration of members of the Surge Capacity Force.

---

[88] On October 6, 2017 the Acting Secretary of DHS extended the deployment cap for voluntary extensions for 45 additional days and encouraged volunteers to speak with their supervisors regarding deployment durations.

California Wildfires

FEMA officials said that although the Surge Capacity Force staff were well trained in individual assistance, they were not always capable of leading teams in austere environments.

Source: FEMA. | GAO-18-472.

Figure 24. Federal Emergency Management Agency (FEMA) Surge Capacity Force Volunteer in Florida.

Concerns Raised About Reservists without Qualified Titles and Who Refused to Deploy

Reservists are FEMA first responders who are available on an on-call basis during an emergency or disaster. They must be available to deploy within 24- 48 hours, and will be deployed for 30 or more days.

- Available reservists as of August 2017: 6,708
- Peak deployed during 2017 disasters: 4,645

As of May 2017, 46 percent of all FEMA reservists did not hold the title of "Qualified" for their job function and from August to November 2017 over 15 percent of eligible reservists refused at least one deployment, according to FEMA data, for medical reasons or other concerns.

Hurricane Harvey (Texas)

Officials from the Joint Field Office—a multiagency coordination center established near a disaster site for coordinating major disaster response and recovery efforts—voiced concern over the deployment of unqualified reservists because training was outdated or unavailable prior to deployment.

Hurricane Irma (Florida)

The FEMA Federal Coordinating Officer in Florida— who coordinates federal activities in support of the state—and Florida Division of Emergency Management officials said morale for some reservists (who made up one-third of staff in Florida during peak deployment) was low due to lengthy deployments, pay cuts, and a shortage of role models. State officials also expressed concern over the number of refusals from deployable reservists who cited medical and scheduling concerns.

Hurricane Maria and Hurricane Irma (U.S. Virgin Islands)

The Federal Coordinating Officer in the U.S. Virgin Islands said many reservists were not physically fit to handle conditions on the island and a fitness test should have been required before they were eligible to deploy.

Concerns with Turnover of Some IMAT Staff and Need for Additional Training

Hurricane Harvey (Texas)

FEMA officials expressed concerns with the high turnover rate of Incident Management Assistance Team (IMAT) staff—the first FEMA emergency management staff deployed to a major disaster site— which they attributed to low pay and the challenging nature of the work.

Hurricane Maria and Hurricane Irma (U.S. Virgin Islands)

FEMA officials noted that attrition is high among IMAT staff and the pay does not incentivize staff to stay. Officials also stressed the need for additional training for Region II IMAT employees alongside national IMAT teams.

APPENDIX VI: FEMA'S INDIVIDUAL ASSISTANCE PROGRAM

Overview

FEMA's IA Program aims to provide individual applicants resources to help meet their sustenance, shelter, and medical needs in the wake of a disaster. FEMA provides substantial assistance through the Individuals and Households Program (IHP)—one of five support programs under IA. The IHP includes two categories of aid:

- Housing Assistance which can include financial or direct assistance for temporary housing, home repairs, replacement of a primary home, and in limited locations, permanent housing construction when needed due to disaster effects.
- Other Needs Assistance which can include financial assistance for uninsured or underinsured, disaster-related needs, such as

transportation, funeral, medical, and child care assistance. Some types of assistance are dependent on an applicant's ability to secure a Small Business Administration disaster loan.

As part of the IHP, individuals affected by disasters declared in fiscal year 2017 may be eligible for up to $33,300 in assistance. Those affected by disasters declared in fiscal year 2018 may be eligible for up to $34,000 in assistance.

In addition to the IHP, FEMA may provide assistance essential to meet immediate threats to life and property resulting from a major disaster, including emergency shelter.

FEMA Individual Assistance Program Activity to Support State Goals for Housing Recovery

The Federal Emergency Management's (FEMA) Individual Assistance (IA) program provides financial assistance and direct services to eligible individuals and households who have uninsured or underinsured necessary expenses and serious needs. In response to the unprecedented 2017 hurricane season, FEMA officials have collaborated with state, territorial and tribal governments to craft new approaches to delivering housing assistance and leverage the broad scope of available authorities under the Stafford Act to meet local needs for housing recovery. These approaches include both financial and direct housing assistance, such as:

- Direct Lease: FEMA provides temporary housing units directly to survivors when rental resources are unavailable. FEMA and the state of Florida are prioritizing use of this approach for housing recovery in the state.
- Multifamily Lease and Repair: FEMA repairs existing multi-family housing units, such as apartments, to use as temporary housing for eligible applicants who are unable to use Rental Assistance—a financial grant provided to homeowners and renters

whose homes were made uninhabitable or inaccessible by the disaster, to assist with expenses to rent temporary housing—due to a lack of available resources. According to FEMA officials, this approach is among the range of options territorial officials intend to leverage to address local conditions.

- Manufactured Housing Units and Recreational Vehicles: These are manufactured homes or other readily fabricated dwellings (e.g., a prefabricated dwelling) owned by FEMA and provided to eligible applicants for use as temporary housing. Recreational Vehicles have been approved for use in response to hurricanes Harvey and Irma. This form of assistance is being implemented in Texas and Florida.
- Permanent Housing Construction: FEMA may provide financial assistance or direct assistance to individuals and households in insular areas outside the continental United States or in other locations where no alternative housing resources are available and where temporary housing assistance is unavailable, infeasible, or not cost-effective. Under this program, repairs can be made to ensure that a home is habitable, such as repairs to heating, ventilation, and air conditioning systems, walls, floors, and ceilings, but is not intended to restore the home to the pre-disaster condition. According to FEMA officials, this assistance was authorized in Texas, Puerto Rico, and the U.S. Virgin Islands.
- Sheltering and Temporary Essential Power (STEP): According to FEMA guidance, STEP was designed to assist state, territorial, and tribal governments in performing work and services essential to saving lives, protecting public health and safety, and protecting property to enable survivors to shelter at home. This approach is among the options state, territorial, and tribal officials told us that they may leverage to address local conditions.
- Transitional Sheltering Assistance (TSA): FEMA may provide TSA to applicants who are unable to return to their pre-disaster primary residence because their home is either uninhabitable or inaccessible. The goal of TSA is to reduce the number of disaster

survivors in congregate shelters by transitioning survivors into short-term accommodations through direct payments to lodging providers, such as hotels. Puerto Rico used this approach despite initial concerns that this would have a negative effect on migration away from the island territory, according to FEMA officials.

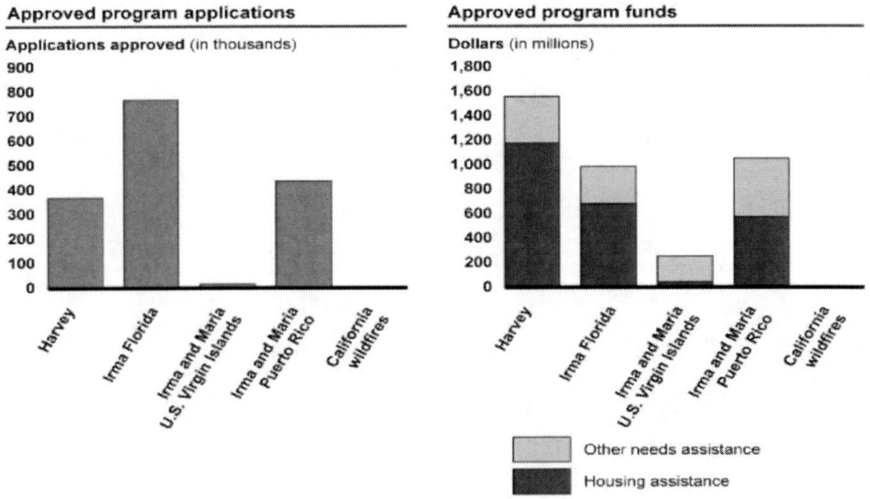

Source: Federal Emergency Management Agency's National Emergency Management Information System. | GAO-18-472.

Figure 25. Total Number of Applicants and Funds Awarded through FEMA's Individuals and Households Program by 2017 Hurricane or Wildfire, as of February 28, 2018.

As of February 2018, FEMA approved more than 1.6 million applications for IHP, resulting in obligations of over $2.5 billion for Housing Assistance and $1.1 billion for Other Needs Assistance (i.e., financial assistance for uninsured or underinsured, disaster-related needs, such as medical), as shown in Figure 25 below. See Table 7 for the number of approved applicants for each FEMA housing and sheltering assistance type for the 2017 disasters as of June 2018.

Table 7. Number of Individual and Households Program Registrations and Approved Applicants for Each Type of Housing and Sheltering Assistance Provided by Disaster Location, as of June 22, 2018

Disaster	Total Registrations	Multifamily Lease Repair	Manufactured Housing Units and Recreational Vehicles	Permanent Housing Construction Repair Program	Direct Lease	Sheltering and Temporary Essential Power Program	Transitional Sheltering Assistance
Hurricane Harvey in Texas	895,528	0	2,848	247	131	15,578	53,894
Hurricane Irma in Florida	2,644,403	0	257	0	63	129	26,633
Hurricane Irma and Hurricane	1,138,444	16	0	33	237	33,016	6,907
Maria in Puerto Rico							
Hurricane Irma and Hurricane	39,415	0	0	0	61	1,920	0
Maria in U.S. Virgin Islands							
California wildfires	25,425	0	154	0	94	0	618
Total	4,743,215	16	3,259	280	586	50,643	88,052

Source: The Federal Emergency Management Agency. I GAO-18-472.

Note: Total registrations represent the number of survivors who applied for disaster assistance at each disaster location. However, the number of registrations for the Sheltering and Temporary Essential Power Program may be different because the Sheltering and Temporary Essential Power Program is separate from FEMA's Individuals and Households Program and is not included in the registration process. Disaster survivors interested in the Sheltering and Temporary Essential Power Program must first apply for FEMA's Individuals and Households Program assistance and then apply directly to the State, Territorial, or Tribal entity administering the Sheltering and Temporary Essential Power Program.

Observations on Individual Assistance Program from Affected States and Territories

Texas – Hurricane Harvey

FEMA entered into an agreement with the Texas General Land Office to provide for housing recovery, marking the first time the agency has coordinated with a non-federal agency to provide this housing service, according to FEMA officials. State officials in Texas will be implementing the new housing approach to manage the delivery of direct housing to more than 6,600 applicants whom FEMA has determined are eligible for direct assistance. FEMA has approved the following direct housing options for eligible applicants in Texas: Permanent Housing Construction Repairs, Direct Lease, Multifamily Lease and Repair, Manufactured Housing Units, and Recreational Vehicles. Local officials cited several advantages to the new housing approach such as the ability to keep homeowners and families in their district thereby sustaining a jurisdiction's tax revenue, supporting businesses, and maintaining public education funding.

Florida – Hurricane Irma

In Florida, state officials preferred to utilize the Direct Lease option to leverage the high volume of vacation rentals, particularly in Lee, Collier, and Monroe counties. FEMA authorized the use of recreational vehicles, purchased directly from commercial dealers, for use as temporary housing, in addition to manufactured homes, as shown in Figure 26. FEMA officials estimated that manufactured homes can cost up to $113,000 while the travel trailers are about $60,000. According to FEMA officials, the decision whether to use the manufactured home, travel trailer or Direct Lease housing options depends on the availability of feasible sites. A FEMA official visits each resident in FEMA housing monthly to check on their progress in transitioning out to their own housing, according to FEMA officials.

Source: GAO. | GAO-18-472.

Figure 26. Example of an Off-the-lot FEMA-purchased Trailer in Key Largo, Florida for Mobility-impaired Survivors.

Puerto Rico – Hurricane Irma and Hurricane Maria

In Puerto Rico, all five IA programs are approved—the Individuals and Households Program, Crisis Counseling, Disaster Legal Services, Disaster Case Management, and Disaster Unemployment Assistance. As of December 2017, more than 1 million residents had applied for IA and FEMA officials extended the deadline to apply through March 2018. FEMA officials told us that they anticipated needing to assist residents in taking next steps to follow up on their applications and collect funds.

According to FEMA officials, they are also using multiple programs and authorities to provide aid to residents with housing needs, including the Direct Lease and the Multifamily Lease and Repair Programs, and the Permanent Housing Construction Repair Program. FEMA is also providing sheltering and emergency assistance through the TSA program, and the STEP program (known as Tu Hogar Renace - Your Home Reborn in Puerto Rico), among others. For example, under STEP, repairs can be done while homeowners remain in place, and the program may provide up to

$20,000 for repairs (although the Federal Coordinating Officer—the lead federal official in charge of response—has the discretion to approve greater costs to accommodate a household's access and functional needs or when the home requires a generator). FEMA estimates that STEP assistance may be provided for 80,000 homes or more. As of June 22, 2018, FEMA had approved 33,016 survivors to use the program in Puerto Rico.

U.S. Virgin Islands – Hurricane Irma and Hurricane Maria

Local officials said they plan to address a shortage of housing through unique housing routes—for example, structured tents, which are used in military operations and are built to withstand 140 mph winds.

California Wildfires

According to California officials, in partnership with FEMA, they established a Housing Task Force to determine how to provide direct housing assistance to thousands of applicants in a timely manner, and incorporate lessons learned from prior wildfire disaster experiences in the state. The task force is also examining options to convert campgrounds into temporary housing.

Appendix VII: Fraud Risk Management in FEMA's Disaster Assistance Programs

Overview

Effective fraud risk management, including controls to prevent, detect, and respond to fraud, can help ensure that federal disaster assistance programs serve their intended purpose, taxpayer dollars are spent effectively, and government assets are safeguarded.

Our Fraud Risk Framework provides a guide for federal program managers to use when developing or enhancing efforts to combat fraud in a strategic, risk-based manner. The framework includes leading practices in four components: (1) Commit to combating fraud; (2) Assess fraud risks; (3) Design and implement a strategy with specific control activities; and (4) Evaluate and adapt fraud risk management activities.

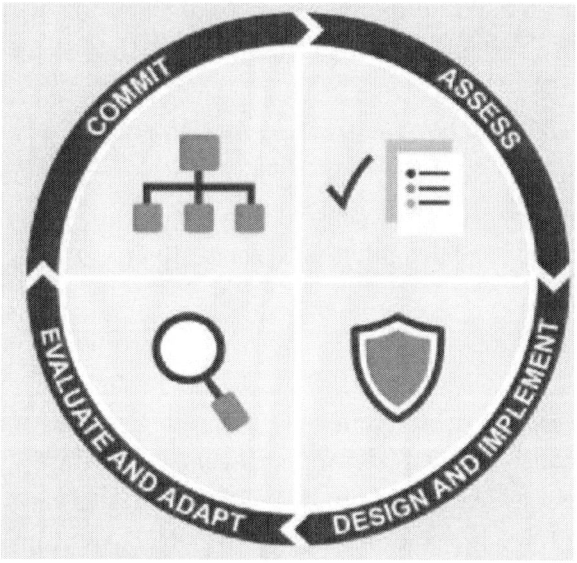

Source: GAO. | GAO-18-472.

Further, the Fraud Reduction and Data Analytics Act of 2015 requires agencies to establish financial and administrative controls that incorporate the Fraud Risk Framework's leading practices.

Observations from Prior Work and Affected States and Territories

The size and scope of the 2017 hurricanes and California wildfires raises questions about the ability of Federal Emergency Management Agency (FEMA) program managers to balance the need to quickly deliver benefits and services while minimizing the risk of fraud, waste, and abuse. Balancing these goals is particularly important given the amount of funds involved—as of June 2018, at least $120 billion had been appropriated to the Department of Homeland Security (DHS), including FEMA, and 18 other federal agencies for activities related to the 2017 disasters.[89]

Since the mid-2000s, FEMA has taken some steps to address identified fraud risks in its disaster assistance programs. Specifically, our prior work found that FEMA strengthened its fraud prevention controls and took other actions to address fraud risks in the Individuals and Households Program (IHP)—one component of the Individual Assistance program.[90] As a result, we identified about $39 million (2.7 percent) of IHP assistance provided following Hurricane Sandy in 2012 that was at risk of being improper or fraudulent, compared to between $600 million and $1.4 billion (10 to 22 percent) of similar assistance provided following hurricanes Katrina and Rita in 2005.

However, the 2017 disasters highlighted the challenges FEMA may continue to face with fraud risks. According to FEMA officials, FEMA identified a well-organized and coordinated identity theft fraud scheme that affected Texas, Florida, Puerto Rico, the U.S. Virgin Islands, and California—a scheme it had not identified following prior disasters. Further, officials from one county we visited as part of this review expressed concern about the risk of fraud and the county's ability to handle disaster payment activities given the volume of transactions the county expects. Moreover, we have previously reported that changes within a program—such as changes in the implementation of the Individual Assistance and Public Assistance programs in areas impacted by the 2017 disasters—can affect the extent to which controls continue to be effective or appropriate for addressing fraud risks.[91]

As described below, FEMA took some steps to address identified fraud risks following the 2017 disasters and earlier events. We are continuing to

---

[89] Supplemental Appropriations for Disaster Relief Requirements Act, 2017, Pub. L. No. 115-56, div. B, 131 Stat. 1129, 1136 (2017); Additional Supplemental Appropriations for Disaster Relief Requirements Act, 2017 Pub. L. No. 115-72, div. A, 131 Stat. 1224, 1224 (2017); Further Additional Supplemental Appropriations for Disaster Relief Requirements Act, 2018, Pub. L. No. 115-123, div. B, subdiv. 1, 132 Stat. 64, 65 (2018).

[90] GAO, *Hurricane Sandy: FEMA Has Improved Disaster Aid Verification but Could Act to Further Limit Improper Assistance*, GAO-15-15 (Washington, D.C.: Dec. 12, 2014).

[91] GAO, *A Framework for Managing Fraud Risks in Federal Programs*, GAO-15-593SP (Washington, D.C.: July 28, 2015).

assess the extent to which FEMA's actions to manage fraud risks in the Public Assistance program align with leading practices described in our Fraud Risk Framework.[92]

Commit: Commit to combating fraud by creating an organizational culture and structure conducive to fraud risk management.

The first component of the Fraud Risk Framework calls for agencies to, among other things, designate an entity to design and oversee fraud risk management activities and to involve all levels of the agency in setting an antifraud tone. Multiple entities within FEMA have designated responsibilities related to fraud risk management. Specifically, the Fraud and Internal Investigations Division (FIID), established in response to fraud associated with major hurricanes in the mid-2000s, is responsible for identifying, mitigating, and preventing fraudulent losses of federal funds and assets by, among other things, reviewing FEMA programs to identify potential improvements to internal controls to prevent and detect fraud, waste, and abuse. In addition, the Director of Risk Management and Compliance is responsible for assisting FIID in the development of antifraud controls. FEMA's Fraud Prevention and Investigation Directive, signed by the Administrator of FEMA in 2014, establishes antifraud responsibilities for several other entities within FEMA, including the Administrator, Regional Administrators, and the Chief Financial Officer, among others. Further, the directive requires all FEMA employees, contractors, and other personnel to take all necessary and proper actions to eliminate fraud, waste, and abuse in FEMA programs.

FIID provides detailed fraud awareness and prevention training—a key responsibility of antifraud entities, according to the Fraud Risk Framework— to FEMA staff, including those responsible for processing disaster assistance applications, according to FIID officials. Increasing

[92] GAO-15-593SP.

employees' awareness of potential fraud schemes—by providing training to stakeholders responsible for program implementation—can help create a culture of integrity and help enable employees to better detect potential fraud.

For the 2017 disasters, FIID added fraud alerts and updates to the daily, pre-shift briefings provided to FEMA IHP intake personnel. These briefings communicated information that the Fraud Risk Framework identifies as key to effective antifraud training, including how to report suspicions of fraud, waste, and abuse; procedures registration and intake personnel and processing staff should follow if an application involves possible fraud or a high-risk applicant; and information on emerging fraud risks and trends identified during the 2017 disasters.

	Assess: Plan regular fraud risk assessments and assess risks to determine a fraud risk profile.

The second component of the Fraud Risk Framework calls for federal managers to identify and assess risks, examine the suitability of existing fraud controls, document a fraud risk profile, and prioritize and determine responses to remaining risks. FEMA has taken some actions to identify and assess fraud risks related to the 2017 disasters. After identifying the identity theft fraud scheme following the 2017 disasters, FEMA hired a contractor in December 2017 to identify and assess fraud risks to the Individual Assistance program, including identifying FEMA stakeholders' fraud risk tolerance and developing a fraud risk profile. In addition, the tasks listed in the contract include, among other things, using the fraud risk profile to assess FEMA's existing controls, reviewing data from past incidents of fraud to identify any control gaps and deficiencies, and making recommendations on ways to improve or add controls. The estimated completion date for the contracted work is August 2018. In addition to the contracted work, FIID is responsible for independently reviewing FEMA programs to identify potential improvements to internal controls to prevent and detect fraud, waste, and abuse. FIID has conducted four reviews of the IHP since 2014 to determine if any indications of fraud are associated with applicants' case files. These reviews did not identify any needed program improvements, according to a FIID official.

FEMA does not have plans to award a contract to identify and assess fraud risks in its Public Assistance program, according to FEMA officials. Instead, according to FEMA's monitoring plan, FEMA incorporates consideration of fraud risk as part of its monitoring approach for Public Assistance grant recipients and conducts a risk assessment of recipients on a rotating, 2-year schedule. Specifically, to determine a grant recipient's risk level, FEMA considers patterns that may reflect recipient issues, such as a history of irregularities in expenditures, a history of disallowed or inappropriate use of funds, and risk of fraud, waste, and abuse. In addition, FEMA considers other factors, such as audit findings, changes in recipient staff, and the dollar value of the grant. According to FEMA officials, FEMA implemented additional controls in Puerto Rico for the Public Assistance grant program based on the results of its risk assessment of Puerto Rico. The additional controls include a manual drawdown process that requires the territorial government of Puerto Rico to fully substantiate all costs claimed for reimbursement before FEMA will authorize the funds for release. In similar risk assessments, FEMA found that grants provided to the U.S. Virgin Islands for recovery from hurricanes Irma and Maria were medium-to-high risk. As a result, FEMA determined it would conduct an onsite monitoring visit for these grants in 2018.

Design and Implement: Design and implement a strategy with specific control activities to mitigate assessed fraud risks and collaborate to help ensure effective implementation.

The third component of the Fraud Risk Framework calls for federal managers to design, implement, and document an antifraud strategy with specific control activities—including reporting mechanisms, data-analytics activities, and fraud-awareness initiatives, among others—to mitigate assessed fraud risks. Further, the Fraud Risk Framework identifies the consideration of the benefits and costs of control activities to address fraud risks as a leading practice, as it can help managers determine if the benefits of a control activity exceed its costs—such as delays for legitimate applicants. During the 2017 disasters, FEMA took steps to design and implement the following antifraud controls, among others:

- Reporting mechanisms: Members of the public and FEMA staff have multiple options to report potential fraud, and FEMA publicized these reporting mechanisms following the 2017 disasters in several ways. For example, FEMA's webpages related to Hurricane Harvey in Texas and Hurricane Irma in Florida include examples of fraud schemes, how to report fraud by phone or email to the National Center for Disaster Fraud, and a link to the fraud website for the DHS Office of the Inspector General fraud, waste, and abuse hotline. We also observed disaster fraud hotline posters on display in the Joint Field Office in San Juan, Puerto Rico. The posters—in both English and Spanish—included multiple options for reporting potential fraud, as shown by the example in Figure 27. The Fraud Risk Framework identifies establishing reporting mechanisms, including hotlines and other mechanisms for receiving tips, as a leading practice for managing fraud risks. Further, the Fraud Risk Framework notes that it is a leading practice for managers to provide multiple options for potential reporters of fraud to communicate and to promote the existence of reporting mechanisms.
- Data-analytics activities: According to the DHS Fiscal Year 2017 Agency Financial Report, FIID conducts data mining of FEMA's databases to identify IHP applications containing common indicators of fraud. The Fraud Risk Framework identifies the implementation of data-analytics activities, including data mining to identify red flags that may indicate suspicious activity, as a leading practice for detecting potential fraud. According to FIID officials, FIID uses data-mining queries to identify red flags, such

as indicators that a damaged dwelling may not be the applicant's primary residence, or instances in which the same Social Security Number was used for different damaged dwellings.

After FIID officials became aware of the identity theft fraud scheme, FIID ran new queries for the 2017 disasters to flag applications with indicators of this scheme, such as applications with questionable banking information, and subjected these cases to additional validation, according to FEMA officials. About 30 percent of Individual Assistance applications in California were flagged as potentially fraudulent because they matched at least one of these queries, which caused a delay in the distribution of funds to actual survivors, according to FEMA officials. According to a FEMA official, the process appears to have been effective at stopping payment on fraudulent applications, although not all of the identified applications were necessarily fraudulent as they may have met one of the flags for legitimate reasons. FEMA will need to assess the process, including the delay for legitimate applicants, according to the FEMA official. According to the contract to identify and assess fraud risks to the Individual Assistance program, the contractor is to develop an implementation strategy, including a cost-benefit analysis, for recommended improvements to existing controls or development of additional controls.

Further, FEMA has taken steps to improve its antifraud controls to prevent and detect fraud and improper payments in response to our findings and recommendations from prior reports. For example, in 2008 FEMA began requiring, among other things, that an inspector meet with an applicant to verify occupancy and confirm that a property was damaged after we identified instances in which FEMA made IHP payments to applicants who used ineligible or bogus addresses following hurricanes Katrina and Rita in 2005.[93]

---

[93] GAO, *Hurricanes Katrina and Rita: Unprecedented Challenges Exposed the Individuals and Households Program to Fraud and Abuse; Actions Needed to Reduce Such Problems in Future*, GAO-06-1013 (Washington, D.C.: Sept. 27, 2006).

Source: National Center for Disaster Fraud. | GAO-18-472.

Figure 27. Examples of Disaster Fraud Hotline Posters Displayed in the Joint Field Office in San Juan, Puerto Rico.

Also, after we determined that FEMA made nearly $17 million in potentially fraudulent or otherwise improper rental assistance payments to individuals through the IHP after they had moved into FEMA trailers following hurricanes Katrina and Rita, we recommended that FEMA take steps to address the issue.[94] In January 2010, FEMA addressed this recommendation by upgrading its data system to display all housing assistance an applicant had received, improving the ability of FEMA caseworkers to identify potential overlapping assistance.

In addition to designing antifraud control activities, collaboration with stakeholders is essential to help ensure that antifraud controls are implemented effectively, according to the Fraud Risk Framework. Specifically, the framework notes that a leading practice for establishing collaborative relationships is to provide guidance and other support to help external parties, such as state and local officials, effectively carry out fraud risk management activities.

---

[94] GAO, *Hurricanes Katrina and Rita Disaster Relief: Continued Findings of Fraud, Waste, and Abuse*, GAO-07-300 (Washington, D.C.: Mar. 15, 2007).

According to FEMA officials, FEMA offers technical assistance, such as programmatic eligibility reviews, to help prevent fraud and established the Procurement Disaster Assistance Team in 2014 to help increase grant compliance among recipients and subrecipients. The Procurement Disaster Assistance Team is tasked with proactively developing and providing training and guidance materials. Officials from one county in California receiving Public Assistance grant funds that we interviewed noted that they received training from FEMA on fraud issues, although officials from another county we interviewed indicated that additional training could be helpful.

Evaluate and Adapt: Evaluate outcomes using a risk-based approach and adapt activities to improve fraud risk management.

Finally, the Fraud Risk Framework calls for agencies to evaluate outcomes using a risk-based approach and adapt activities to improve fraud risk management. FEMA has taken steps to measure outcomes of its antifraud approach for the IHP. FIID evaluates the success of its antifraud approach for the IHP—which has shifted from a reactive to a preventative model, according to FEMA officials—by comparing the amount of potentially fraudulent funds it has prevented from being disbursed with what is submitted for recoupment. Specifically, according to the DHS Fiscal Year 2017 Agency Financial Report, FIID locks IHP applicant files that contain common indicators of fraud in order to prevent fraudulent funds from being disbursed. As a result, according to the report, FIID prevented $20.6 million from being disbursed in fiscal year 2017 and submitted $3.6 million for recoupment, compared with fiscal year 2014 when it prevented $4.4 million from being disbursed and submitted $7.2 million for recoupment.

In addition, FIID takes steps to adapt its antifraud data analytics for the IHP. According to FEMA officials, after FIID identified the identity theft fraud scheme following the 2017 disasters, FEMA suspended emergency payments for critical needs to over 200,000 suspicious applicants in Texas, Florida, Georgia, Puerto Rico, and the U.S. Virgin Islands, which prevented millions of dollars in potentially fraudulent disaster payments from being disbursed. FIID also evaluates the success of its data-mining queries by calculating the percentage of IHP applicants in the query that were found to be fraudulent. FIID then implements the most effective queries for all disasters, according to FEMA officials. Further, according to FEMA program officials, they evaluate the effectiveness of internal controls through testing to identify improper payments, quality control reviews, and audits.

Appendix VIII: Payment Integrity and Prior Identified Requirements for Disaster Relief Funding

Overview

When disasters occur, the destruction they cause must be addressed immediately, and disaster relief funding must be delivered expeditiously. However, the risk for improper payments increases when billions of dollars are being spent quickly. For many years, GAO and the Inspector General community have identified internal control weaknesses in the federal government related to agencies receiving supplemental funds for disaster assistance.

Standards for Internal Control in the Federal Government sets the standards for an effective internal control system for federal agencies and provides the overall framework for designing, implementing, and operating an effective internal control system.

Mandated Requirements to Ensure Payment Integrity for the 2017 Disasters

With supplemental appropriations totaling at least $120 billion in additional funding for activities related to the 2017 hurricanes and wildfire disasters, Congress provided an oversight framework for these funds related to internal control and improper payments. Congress included the following key payment integrity provisions to help assure that all the funds are being spent as efficiently and effectively as possible:[95]

- Federal agencies are required to submit their plans for ensuring internal control over disaster relief funding to GAO, respective Inspectors General, the Office of Management and Budget (OMB), and Congress.
- OMB is required to issue standard guidance for federal agencies to use in designing internal control plans for disaster relief funding.

Observations from Prior Work on Payment Integrity

We have previously reported deficiencies related to federal agencies' establishment of required internal control plans in response to natural disasters and OMB's guidance for development of those plans. Specifically, in 2013, we reported on deficiencies in the internal control plans related to Hurricane Sandy disaster funding. These concerns may continue to be an issue for agencies after the 2017 disasters and we will monitor their efforts as part of our ongoing work.

Federal Agencies' Internal Control Plans

Agencies prepared Hurricane Sandy disaster relief internal control plans based on OMB guidance but did not consistently apply the guidance

[95] See Pub. L. No. 115-123, § 21208, 131 Stat. 64 (2018); Pub. L. No. 115-72, § 305, 131 Stat. 1224, 1227-28 (2017).

in preparing these plans. OMB Memorandum M-13-07, Accountability for Funds Provided by the Disaster Relief Appropriations Act, directed federal agencies to describe incremental risks they identified for Hurricane Sandy disaster relief funding and provide internal control strategies for mitigating these risks. Each of the 19 agencies responsible for the 61 programs receiving funds under the Disaster Appropriations Act 2013 submitted an internal control plan with specific program details using a template that OMB provided. In November 2013, we reported that agencies' plans ranged from providing most of the required information to not providing any information on certain programs. For example, each of the 61 programs was required to discuss its protocol for improper payments; however, we found that 38 programs included this information, 11 included partial information, and 12 included no information.

OMB Guidance for Development of Internal Control Plans

We also reported that OMB's guidance was an important step in the oversight of Hurricane Sandy disaster funding, addressing internal controls, improper payments protocol, and unexpended grant funds. However, we identified several weaknesses in OMB's guidance that limited its effectiveness in providing a comprehensive oversight mechanism for these funds. Specifically, the guidance (1) focused on identifying incremental risks without demonstrating that known risks had been adequately addressed; (2) provided agencies with significant flexibility as it did not require documentation or criteria for claiming exceptions, such as why the OMB requirements were not feasible or practicable; and (3) resulted in certain agencies developing their internal control plans at the same time that funds needed to be quickly distributed. We recommended that OMB develop more robust guidance for agencies to design internal control plans for future disaster relief funding. In commenting on the draft report, OMB staff generally agreed with our recommendation. On July 15, 2016, OMB issued the revised Circular No. A-123, *Management's Responsibility for Enterprise Risk Management and Internal Control*. The circular requires agencies to implement enterprise risk management, which includes developing a risk profile that analyzes the risks faced in achieving strategic objectives and identifies options for addressing them. However, the revised circular did not include specific guidance for identifying risks related to disaster funding; thus, the recommendation remains open. We plan to continue monitoring OMB's progress in implementing this recommendation.

Appendix IX: FEMA's Public Assistance Grant Program

Overview

FEMA's Public Assistance grant program is administered through a partnership between FEMA and the state grantee, which provides funding to local officials.

In recent years, FEMA has redesigned the PA program to address past challenges and make the program easier for FEMA and grantee officials to manage. These efforts represent FEMA's "new delivery model" for awarding PA program grants. Officials implemented the new model in Texas and Florida after hurricanes Harvey and Irma, and announced the national use of the new model for PA concurrent with this decision.

Public Assistance Program Activities for Disaster Recovery

As of February 2018, FEMA had obligated close to $1.5 billion in PA grants to three states and two territories recovering from hurricanes Harvey, Irma, and Maria, as well as the California wildfires for emergency work projects— such as debris removal. For the longer-term projects—such as rebuilding of public facilities and infrastructure it may take months, and in some cases years, to award grant funds to state and local governments to aid in their disaster recovery—FEMA had obligated close to $2 billion as of February 2018. See Figure 28 for the PA obligations for emergency and permanent projects for the 2017 disasters.

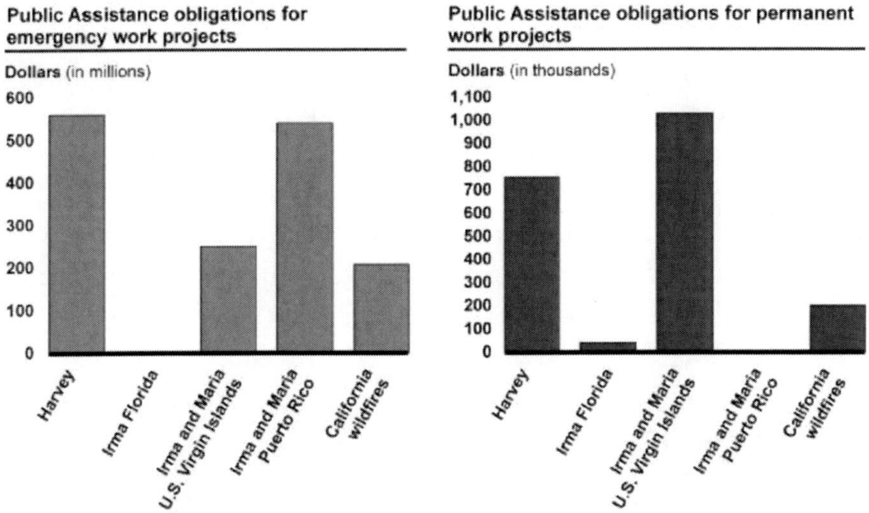

Source: Federal Emergency Management Agency's Emergency Management Mission Integrated Environment. | GAO-18-472.

Note: Emergency work projects include debris removal and emergency protective measures, such as flood fighting activities, evacuation and sheltering of disaster survivors, and providing medical care and transport. Permanent work projects include the repair of damaged facilities such as the repair of dirt roads.

Figure 28. Public Assistance Obligations for Emergency and Permanent Work for Each of the 2017 Disasters, as of February 2018.

FEMA and local officials we interviewed in Texas, Florida, Puerto Rico, and the U.S. Virgin Islands noted early recovery challenges, in part, related to the new approaches to PA grant administration. Specifically, officials identified:

Debris Removal Challenges

Local officials managing the recovery from each disaster raised concerns about challenges with debris removal, for example, such as the limitations of local officials' pre-positioned contracts that slowed the pace of removal.

PA New Delivery Model

Texas, Florida, and California are using the new PA model. Under the new model, FEMA redesigned processes for developing, reviewing, and approving grant applications. The agency created specialized roles; implemented a centralized and standardized grant processing approach; took steps to increase hazard mitigation; and is developing a new information system to better maintain and share grant documentation. However, all three states have faced challenges with implementation due to lack of experience and expertise.

PA Alternative Procedures

FEMA began the pilot approach after Hurricane Sandy, to provide more flexibility for program administration and incentivize applicants to complete projects in a timely and cost-effective manner. Puerto Rico will use the PA Alternative Procedures; however, the process is under development in Puerto Rico and FEMA has not used it at this scale before.

Observations from the 2017 Disaster-Affected States

Experiences with the PA program in the affected states varied across the four 2017 disasters, but early observations from officials indicate that challenges with debris removal, the new PA delivery model, and the PA Alternative Procedures may result in long-term effects on community and infrastructure recovery.

Debris Removal Challenges in Each Disaster

- *Hurricane Harvey (Texas)*: Local officials in Texas noted that debris removal was a challenge, but indicated that they were able to make progress through internal efforts and coordination with FEMA. For example, local officials in one jurisdiction noted problems with the use of pre-positioned contracts. In this case, the officials hired military personnel to help move debris. In another case, local officials noted delays in debris removal due to resource constraints—having only 25 trucks—and receiving slow responses from FEMA, on questions such as how to address debris removal from gated communities. Local officials that did not note concerns with debris removal credited coordination with FEMA officials, expedited funding from FEMA, and daily calls with the state as the basis for their success.

- *Hurricane Irma (Florida)*: Local officials in Florida highlighted debris removal as the greatest challenge they experienced early in the recovery. The Federal Coordinating Officer (FCO)—the lead federal official in charge of response—in Florida reported that there was a shortage of contractors for debris removal because contractors are primarily located in the Gulf Coast area and serve a national market. See Figure 29 for a picture of debris that was awaiting pick up by contractors in Big Pine Key, Florida. Also, many debris contractors were still engaged in Texas following Hurricane Harvey. Further, some preexisting contracts were awarded up to 5 years before Hurricane Irma, but the market had changed and newer contracts were offering more money per cubic yard of debris removal. As such, some vendors, despite being under contract in other locations, prioritized work in jurisdictions that offered higher rates according to the FCO in Florida. According to officials in one county, they had to compete for debris removal contractors, after contractors increased their prices, due to the high demand across Texas and other parts of Florida in the aftermath of the hurricanes. Local officials in Florida also faced challenges identifying who was responsible for debris removal from waterways, raising concerns about the environmental impacts.

Source: GAO. | GAO-15-472.

Figure 29. Staging Debris along Route 1 near Big Pine Key, Florida October 2017.

- *Hurricane Irma and Hurricane Maria (U.S. Virgin Islands)*: Officials faced unique challenges with staging and debris removal, due to the widespread vegetative damage across the islands. According to a FEMA official, there were challenges supporting the debris removal operation because local officials did not want to burn the vegetative debris.
- *Hurricane Irma and Hurricane Maria (Puerto Rico)*: According to FEMA and local officials in Puerto Rico, debris removal will require coordination among local contractors hired by each municipality, the U.S. Army Corps of Engineers, and other agencies supporting recovery. According to Puerto Rico officials, local officials may experience challenges with reimbursement for debris removal activities. Further, the officials said there were resource constraints, so they had to prioritize debris removal from state-managed roads, before clearing local roads.

- *California Wildfires*: In northern California alone, the wildfires created the largest amount of debris since the 1906 earthquake, and state officials and contractors have almost completed efforts to remove debris as of March 2018, according to state officials. In California, the debris removal contract with the U.S. Army Corps of Engineers created recovery challenges due to the lack of flexibility in contracting requirements, according to local officials. Furthermore, according to California officials, the contract process resulted in a bid protest that would have delayed debris removal, but FEMA and state officials worked together to address these issues and prevent delays.

Local Experiences with the New PA Delivery Model

- *Hurricane Harvey (Texas)*: Local officials' early recovery experiences with the PA program varied. Generally, local officials noted the potential for the new process or information system to improve PA grant delivery. However, local officials also noted a lack of consistency in eligibility determinations and the knowledge and experience of program staff as presenting a potential challenge for their overall recovery. For example, officials in one county raised concerns about inconsistent eligibility determinations, where there are still open disaster declarations and PA projects from flooding events in 2015 and 2016 that occurred prior to Hurricane Harvey. FEMA had not obligated funds for these projects, due to changing eligibility determinations made by FEMA officials throughout the process, which may subsequently impact obligations for Hurricane Harvey projects, according to county officials. However, FEMA headquarters officials noted that the new delivery model has not changed eligibility criteria or authorities for eligibility determinations and was not being applied retroactively to old disasters. In another county, officials also noted that FEMA staff could not answer questions on the new PA process and this raised concerns that PA staff were not documenting information in the new IT system. In particular, the officials cited challenges with support for developing mitigation proposals, and hired contractors to help develop mitigation project proposals after hazard mitigation specialists were unable to provide assistance. Moreover, the officials stated that they did not receive training on the new model until after Hurricane Harvey. Other local officials also noted similar gaps in FEMA-provided training and said they lacked enough skilled staff to support their projects.

- *Hurricane Irma (Florida)*: According to local officials, problems with FEMA customer service raised concerns about the long-term effect on recovery efforts. Local officials noted that PA staff did not have the knowledge or experience necessary to provide the support they needed throughout the process. For example, in one county PA staff provided assistance learning how to use the new IT system, but had not provided the detailed training on what types of information are required. In contrast, another local county official noted positive experiences with PA staff, and cited the potential for the new delivery model and new IT system to improve the program operations while reducing the administrative costs.
- *California Wildfires*: According to state officials, wildfires cause more damage to individual property owners, and therefore the public infrastructure recovery will be limited and the majority of PA projects will be emergency protective measures and debris removal. According to officials in one county, they had prior experience with the new model and process and received quick support when questions arose on the new IT system. Therefore, there have not been any problems or delays processing projects, according to local officials. The officials added that they have opted not to submit all damages, due to the documentation burden, as it is not worth their time to compile the paperwork for the project based on the amount of the award. In a second, more rural county, officials faced similar challenges but cited a lack of

training for their county, inadequate support from FEMA, and untrained PA staff as challenges for their PA recovery process under the new model.

Use of PA Alternative Procedures Pilot Project

In response to the hurricanes, the governor of Puerto Rico requested the use of the PA Alternative Procedures, which FEMA began piloting after Hurricane Sandy. In Puerto Rico, Alternative Procedures are to be used for all large permanent work projects, FEMA, in collaboration with local officials, issued guidance specific to the Hurricane Maria recovery in Puerto Rico in April 2018. According to FEMA officials, using these flexibilities is unlikely to result in a faster recovery, but may offer more opportunities to rebuild with greater resilience from future hurricanes and natural hazards. FEMA officials also noted that capacity limitations—both in the PA workforce and in local officials' experiences with the program—have hampered early recovery efforts. According to FEMA officials, they did not use the new model in Puerto Rico specific to the Hurricane Irma recovery because PA program officials released their new model guidance after Hurricane Irma recovery efforts were already underway, and officials did not want to change the process again after Hurricane Maria.

In all other locations, state and local officials have the option to use the PA Alternative Procedures. For example, in the U. S. Virgin Islands, local officials are using the Alternative Procedures pilot to meet their needs. FEMA officials in the U. S. Virgin Islands noted similar challenges with a lack of PA funding and territorial personnel to develop PA projects, as well as limitations in the skill level for those PA staff assigned to the islands. According to FEMA officials, the U.S. Virgin Islands hired an emergency management contractor to assist in preparation and oversight of PA projects to address the staffing challenges that exist for FEMA and local officials. FEMA officials also said that complex projects, such as a wastewater treatment plan in St. Thomas and two major hospitals, may present fiscal challenges in the long term.

APPENDIX X: DISASTER RESILIENCE AND HAZARD MITIGATION

Overview

The costs of severe weather events, such as those seen in 2017, are likely to continue to rise due to climate change. We have reported that, enhancing disaster resilience and hazard mitigation is essential for controlling federal fiscal exposure to disasters. However, the current funding approach, which emphasizes the post-disaster environment, can create a reactionary and fragmented approach where disasters determine when and where investment occurs. A national strategic approach to prioritizing investments could help ensure that federal funds are directed towards the most effective risk reduction efforts.

The Mitigation Framework Leadership Group—the interagency body responsible for guiding federal hazard mitigation efforts under the National Preparedness System—has released a draft National Mitigation Investment Strategy, which articulates key principles and desired outcomes to help guide a national approach to resilience investments. In addition, FEMA has included new guidance and training as part of its new delivery model for Public Assistance to help ensure that applicants seek opportunities to incorporate hazard mitigation.

Resilience and Hazard Mitigation

Disaster resilience is the ability to prepare and plan for, absorb, recover from, and more successfully adapt to adverse effects. Hazard mitigation actions are undertaken to enhance disaster resilience by reducing or eliminating long-term risk to people and property from natural hazards and their effects.

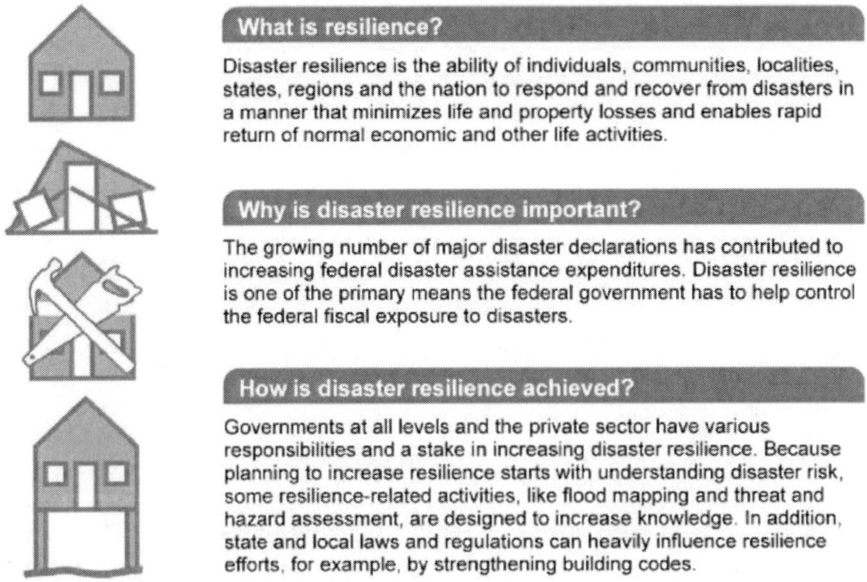

Source: GAO analysis. | GAO-18-472.

Figure 30. Resilience – What, Why and How.

Federal Pre-Disaster and Post-Disaster Hazard Mitigation Programs

The Federal Emergency Management Agency (FEMA) and other federal agencies have multiple funding mechanisms, including those outlined below, to help states and localities enhance disaster resilience and hazard mitigation—the majority of which are available after a disaster strikes.

FEMA's Pre-Disaster Mitigation Program provides funding to state, local, and tribal governments to help plan for and implement hazard mitigation projects prior to a disaster. The program's goal is to reduce overall risk to the population and structures from future hazard events.

FEMA's Hazard Mitigation Grant Program provides funding to protect states, tribes, and territories after a major disaster is declared. The recipient can then use the funds for eligible projects anywhere in the state, tribe, or territory to reduce the risk of future disaster damage.

FEMA's Public Assistance Program provides funding for hazard mitigation measures to the parts of a facility that were damaged by a disaster. Funding is limited to declared counties and eligible damaged facilities. Mitigation measures can also be applied, in certain circumstances, to non-damaged facilities under the Public Assistance Alternative Procedures—procedures that give jurisdictions more flexibility in determining how, where, and what to rebuild, particularly after incurring significant damage—so long as they are otherwise eligible under the Public Assistance program.

Housing and Urban Development Community Development Block Grant – Disaster Recovery provides funding to address needs not met by other disaster recovery programs post-disaster, which can include disaster resilience-building projects. Funding is provided to affected cities, counties, and states, especially in low-income areas, through congressional supplemental appropriations.

Observations from Affected States and Territories

FEMA and local officials noted how previous mitigation projects lessened the damage from the 2017 disasters in some areas and described some challenges incorporating resilience as well as plans to incorporate resilience moving forward.

Examples of Previous Successes and Challenges with Hazard Mitigation

- *Hurricane Harvey (Texas):* FEMA officials said Hurricane Harvey demonstrated how prior hazard mitigation projects prevented greater damages (e.g., elevated homes and equipment sustained less damages as shown in Figure 31).

- *Hurricane Irma (Florida)*: FEMA officials said Florida strengthened its building codes for resilience as a result of lessons learned from Hurricanes Andrew in 1992 and Mathew in 2016 (e.g., elevating homes). State officials noted that some areas did not lose power during Hurricane Irma due to a previous hazard mitigation effort that reinforced power poles as shown in Figure 32.

Source: GAO. | GAO-18-472.

Figure 31. Elevated Air Conditioning Unit in Greater Houston Area, October 2017.

- *Hurricane Irma and Hurricane Maria (Puerto Rico):* Puerto Rico's construction codes required buildings to withstand winds of 145 miles per hour, but the force of Hurricane Maria's winds exceeded that speed at 175 miles per hour resulting in damage, according to FEMA officials.
- *Hurricane Irma and Hurricane Maria (U.S. Virgin Islands):* Previous resilience efforts helped in the U.S. Virgin Islands. For example, hurricanes Irma and Maria destroyed 90 percent of the

power grid; however, due to a hazard mitigation project undertaken as a result of Hurricane Marilyn in 1995, 30 percent of the power grid was able to be restored quickly, according to FEMA officials.

Source: GAO. | GAO-18-472.

Figure 32. Reinforced Power Line Poles in Monroe County, Florida, October 2017—Post Hurricane Irma.

Examples of Plans to Incorporate Resilience Following the 2017 Disasters

- *Hurricane Harvey (Texas):* State officials noted that localities may be more willing to incorporate resilience post-Harvey due to grants from Rebuild Texas—a state-run program intended to make counties more resilient to future storms and flooding. In addition, FEMA officials said they have staff identifying opportunities for mitigation on infrastructure projects.
- *Hurricane Irma (Florida)*: FEMA has held workshops with local residents on hazard mitigation measures such as adding window shutters, strapping down roofs, and installing different types of glass, according to agency officials.
- *Hurricane Irma and Hurricane Maria (Puerto Rico):* Puerto Rico received $18.5 billion from the Housing and Urban Development Community Development Block Grant-Disaster Recovery program. While most will be spent on housing, the governor plans to invest remaining funds in power grids, infrastructure, and other hazard mitigation measures according to Puerto Rico officials.
- *Hurricane Irma and Hurricane Maria (U.S. Virgin Islands):* U.S. Virgin Islands officials are receptive to investing in infrastructure improvements, according to FEMA officials. For example, U.S. Virgin Islands officials formed a task force—including FEMA mitigation and critical infrastructure officials, the Virgin Islands Territorial Emergency Management Agency, the Department of Transportation, and other members of the governors' office—focused on rebuilding power infrastructure and two major hospitals in a resilient way, according to FEMA officials.
- *California Wildfires:* FEMA officials said they have provided wildfire resilience materials at local rebuilding fairs and are working with the state to provide mitigation grant application workshops.

Appendix XI: Department of Defense's Support of Civil Authorities during the 2017 Hurricanes and Wildfires

Overview

While DOD's primary mission is to defend the nation, the department is often asked to play a prominent role supporting civil authorities and must be prepared to provide rapid response when called upon during disasters and declared emergencies (natural or manmade). DOD provides such support through its Defense Support of Civil Authorities mission.

Consistent with the *National Response Framework*—a guide to how the federal government, states and localities, and other public and private-sector institutions should respond to disasters and emergencies—DOD primarily provides support through two approaches: (1) when requested (e.g., mission assigned) by a federal agency (e.g., FEMA, the Department of Health and Human Services, or U.S. Department of Agriculture) and approved by the Secretary of Defense, the department provides federal military forces, DOD civilians, DOD contract personnel, and DOD component assets; and (2) the U.S. Army Corps of Engineers serves as the DOD coordinating and primary federal agency for public works and engineering related response efforts.

Separate from DOD's efforts, National Guard units provide support to their governor or other governors through state-to-state emergency management assistance compact agreements. In this capacity, National Guard units and personnel do not operate as a DOD (federal) capability or resource. Instead, they serve under state law and are funded with state resources— a status commonly referred to as State active-duty status. In addition, when approved by the Secretary of Defense with the concurrence of the affected governor, National Guard units also provide support with DOD funding -- commonly referred to as Title 32 status. In both situations, the National Guard is under the control of their governor.

Summary of DOD's Support for 2017 Disasters

The Department of Defense (DOD) provided extensive support during the 2017 hurricanes and wildfires. This support included routine support (e.g., providing food, water, planners, debris removal, temporary roofing, and federal partners access to DOD bases and facilities for staging response personnel and equipment) as well as capabilities that have not been routinely requested (e.g., using U.S. Navy ships as helicopter platforms, procuring and installing large generators, providing medical support for prolonged period of time, and power grid restoration). As shown in Table 8, DOD components and organizations at all levels provided support:

Table 8. DOD Components and Personnel that Provided Support

Department of Defense Components[a]
• Office of the Deputy Assistant Secretary of Defense for Homeland Defense Integration and Defense Support of Civil Authorities
• Joint Staff
• U.S. Northern Command
• U.S. Transportation Command
• National Guard Bureau
• Defense Logistics Agency
• U.S. Army Corps of Engineers
• Defense Coordinating Officers and Elements
• U.S. bases (e.g., Joint Base San Antonio, Texas, and Fort Benning, Georgia)
• Military units and personnel

Source: DOD I GAO-18-472.

[a] Separate from DOD's Title 10 efforts (and the 164 National Guard members who responded in Title 10 federal status), more than 56,000 Guard members from 49 states, the District of Columbia, the U. S. Virgin Islands, and Puerto Rico provided support in either State Active Duty or Title 32 status for Hurricanes Harvey, Irma, and Maria; and the California Wildfires, according to the National Guard Bureau.

The Federal Emergency Management Agency (FEMA) issued work orders directing DOD to provide support and assistance for the 2017 disasters (mission assigned).

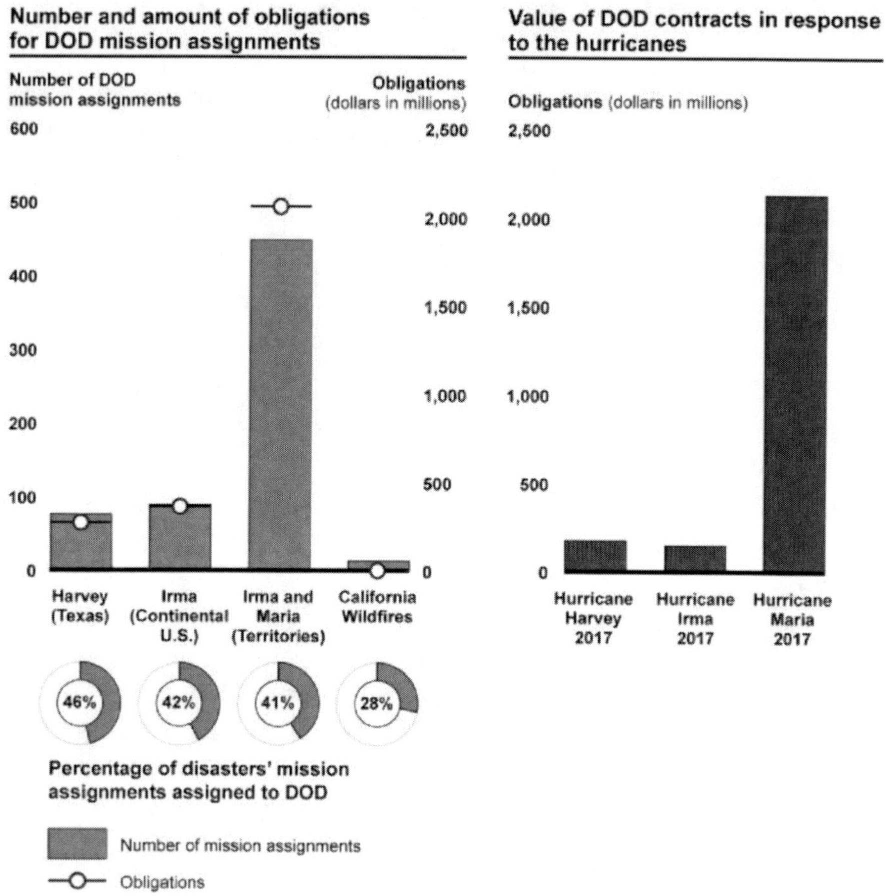

Source: GAO analysis of mission assignment data from the Federal Emergency Management Agency's (FEMA) Web-Based Emergency Operations Center (WebEOC), obligations from FEMA's Integrated Financial Management Information System, and contracting data from the Federal Procurement Data System-Next Generation. | GAO-18-472.

Note: Mission assignment totals and percentages reflect data entered into FEMA's Web-based Emergency Operations Center system of approved resource requests by response officials, where data are available, as of January 2018. This does not include 1,285 request records (out of 3,338 total) for which data are not available to determine whether FEMA sourced the request through a mission assignment or other source, and 74 records for which data are not available to determine which agency FEMA assigned out of the 1,515 mission assignments reviewed.

Figure 33. DOD Mission Assignment and Contract Activity for Four 2017 Disasters.

As shown in Figure 33, the number of mission assignments, cost of mission assignments, and value of contracts issued by DOD varied for each of the 2017 disasters. For example, the estimated cost for a mission assignment that requested Civil Air Patrol assistance was $5,000 while the cost for the U.S. Army Corps of Engineers to restore the Puerto Rico electrical power grid was $577,000,000.

DOD was asked to support more mission assignments in the U.S. Virgin Islands and Puerto Rico than in Texas and Florida due to a number of factors. According to DOD, state and local officials in Texas and Florida were more experienced and prepared to respond to the hurricanes that affected their states and were able to rely on National Guard and other resources from unaffected areas of the state. Also, according to DOD officials, due to the circumstances—including power grid destruction and being accessible only by air or sea—DOD was asked to provide support in the U.S. Virgin Islands and Puerto Rico that the department has not typically provided for prior hurricanes.

Observations from Affected States and Territories

Hurricane Harvey (Texas): While DOD provided support prior to Hurricane Harvey's landfall (e.g., pre-positioned generators and fuel), the majority of mission assignments that FEMA requested from DOD occurred after the storm stalled over southeast Texas. Such support included:

- Army units provided high-water vehicles and Marine Corps units provided amphibious vehicles and boats to rescue over 6,000 citizens from flooded areas in support of Texas and to transport supplies for the American Red Cross and support FEMA logistics efforts, as shown in Figure 34.
- Air Force, Navy, and Army aircraft provided search and rescue aircraft saving over 1,000 individuals. Air Force also provided airborne command and control aircraft and imagery that provided improved capability for numerous interagency aircraft supporting

search and rescue operations to safely operate in very busy airspace.

- Air Force C-130s sprayed over 2.7 million acres of areas that had troublesome mosquito populations.

Source: U.S. Army. | GAO-18-472.

Figure 34. Army High-Water Vehicle use in Texas.

- Joint Base San Antonio provided two separate locations (Randolph AFB and Seguin Auxiliary Airfield) as staging areas for thousands of tractor trailers that were used to distribute commodities. Naval Air Station Corpus Christi provided FEMA the use of a nearby training airfield where tents, showers and feeding facilities could be established to house emergency responders so they did not compete with hurricane survivors for local hotel spaces.

Hurricane Irma (Florida): With DOD providing ongoing support for Hurricane Harvey victims and Hurricane Irma having hit the Caribbean a few days earlier, DOD provided support prior and subsequent to the hurricane hitting Florida. DOD support included:

- U.S. Transportation Command provided emergency response planners and used its aircraft to fly FEMA's urban search and rescue teams, Health and Human Services disaster medical assistance teams, and relief supplies and equipment into the state, as shown in Figure 35.

Source: U.S. Air Force. | GAO-18-472.

Figure 35. Air Force Reservist Prepare a Pallet of Supplies for Shipment to Florida.

- Navy personnel from the USS *New York*, USS *Iwo Jima*, and USS *San Jacinto* provided 6,372 meals, 14,719 gallons of water, and 1 medical evacuation in the vicinity of Key West.
- Special Operations personnel supported FEMA's public affairs efforts.
- DOD provided geo-imagery analysts to FEMA's National Response Coordination Center that assisted FEMA with damage assessment capabilities.

- 10 military installations were used for commodity and equipment staging, FEMA Incident Management Assistance Teams, operations facilities, and Urban Search and Rescue teams' accommodations.

Hurricane Irma and Hurricane Maria (Puerto Rico/U.S. Virgin Islands): DOD was able to reposition U.S. Navy ships that were enroute to support Hurricane Harvey to the U.S. Virgin Islands; these and other DOD capabilities that were providing assistance to the Virgin Islands were able to provide immediate support to Puerto Rico after Hurricane Maria hit the island as shown in Figure 36. DOD support included:

- The Marine Corps provided two Doppler radar units, meteorologists, and radar technicians that provided weather forecasts and facilitated aviation safety on the U. S. Virgin Islands and Puerto Rico.

Source: U.S. Navy. | GAO-18-472.

Figure 36. Supply delivery to U. S. Virgin Islands and Puerto Rico.

- U.S. Army Reserve unit provided mortuary affairs services at multiple locations, including local hospitals.
- DOD supported the State Department in evacuating approximately 6,000 U.S. citizens from the British Virgin Islands to Puerto Rico.
- Army and Air Force set up medical support hospitals and the Navy provided medical services aboard and alongside the USNS *Comfort*. For example, the Army deployed a temporary medical facility to St. Croix that included emergency medical care services, urgent care medical services, temporary patient holding, and ancillary services to triage and medically treat approximately 200 disaster victims per day to stabilize medical care at local hospitals.
- Special operations units provided information support, which included public affairs messaging to the affected population and had a major impact in communicating messages to the affected population when there was little or no radio or TV public broadcasts.
- U.S. Army Corps of Engineers provided temporary emergency power, temporary roofing, debris management, infrastructure assessment, critical public facility restoration and temporary housing. The U.S. Army Corps of Engineers is also repairing the power grid in Puerto Rico.
- Defense Logistics Agency provided, among other things, power poles, pharmaceuticals, handheld radios, generators, water, meals, human remains pouches, and fuel.

2017 California Wildfires: According to DOD officials, National Guard personnel in State Active Duty status primarily provided wildfire fighting capabilities. However, DOD provided commodity support and access to military bases. For example,

- Defense Logistics Agency fulfilled 3,500 orders placed by the U.S. Forest Service for emergency equipment and supplies, including 5 million AA batteries.

- Air Force provided access and use of military bases (including Travis Air Force Base and Beale Air Force Base) to stage ambulances and their crews, as shown in Figure 37, as well as housing support for federal personnel in various agencies.

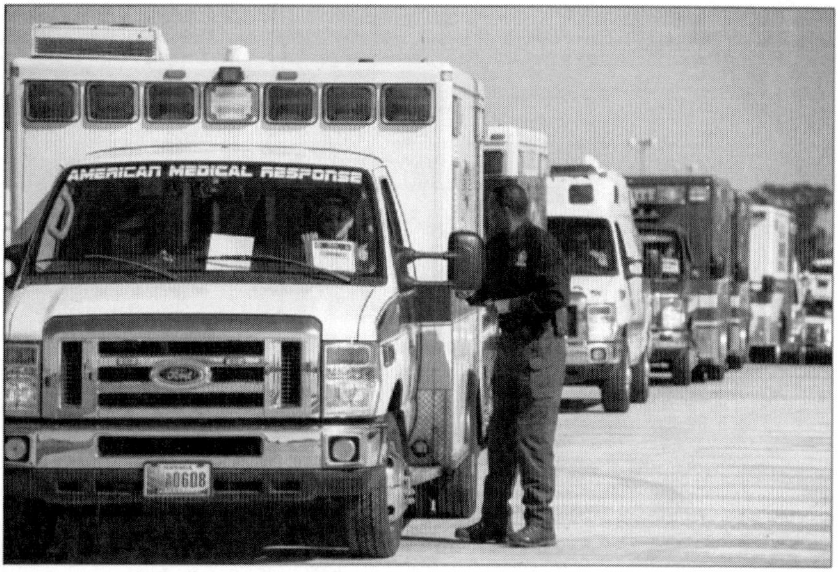

Source: U.S. Air Force. | GAO-18-472.

Figure 37. Ambulance and Support Vehicles Staged at Travis Air Force Base, California.

Challenges and Lessons Learned from 2017 Disasters

DOD officials identified a number of challenges that they encountered during disasters that they recognize need to be addressed by DOD and its federal partners in the future. Such challenges include:

- *FEMA and Emergency Support Function lead agencies' dependence on DOD capabilities.* DOD officials stated that federal agencies have over time become too dependent on DOD capabilities during disasters. While DOD possesses some unique capabilities, some of the requested capabilities could potentially reside in other federal agencies, nongovernmental organizations, or the private sector. Similarly, according to DOD officials, the department's ability to deploy quickly and/or for extended periods of time may make DOD a preferable solution for response capabilities and support. The increased reliance may create vulnerability, if in the future, DOD capabilities are needed to conduct DOD's primary mission—to defend the nation from threats concurrent with a domestic disaster response.

- *DOD units and personnel were deploying without authorization.* DOD officials, including those from the National Guard Bureau and DOD coordinators located in FEMA regions, told us units were activating or taking action too early, which results in units self-deploying without authorization to disaster areas and potentially adversely impacting FEMA-coordinated response efforts in the areas where the units deploy.
- *Potential impact to readiness.* DOD and National Guard efforts to support civil authorities in the hurricanes and wildfires in the United States (as well as provide humanitarian support to other countries and territories) may have affected the readiness of units and commands to conduct global military operations from 2018-2020, according to DOD and National Guard officials.
- *Unclear mission integration.* DOD officials noted that integrated planning for some Emergency Support Functions, such as public health and medical services, prior to the 2017 disasters had not occurred. Such planning could have identified the status of capabilities across the government, including DOD capabilities (e.g., medical command and control elements and medical response elements). This would have clearly defined how federal departments and agencies, including DOD, should be providing support (e.g., patient evacuation and deployable medical treatment facilities).

Appendix XII: Federal Emergency Management Agency (FEMA) Regional Structure

FEMA has 10 Regional offices located across the United States as depicted below in Figure 38.

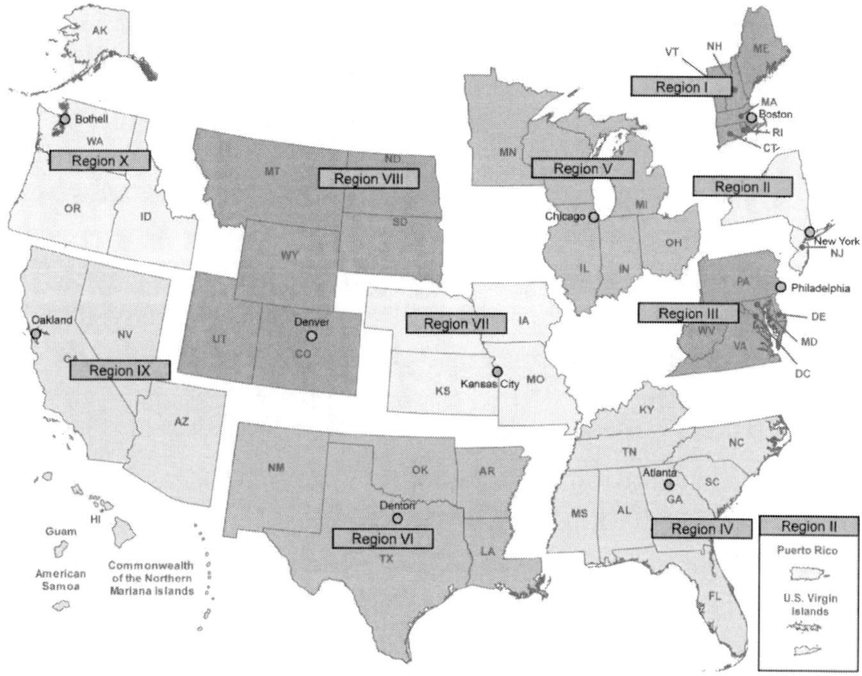

Source: Federal Emergency Management Agency; Map Resources (map). | GAO-18-472.

Figure 38. Map of Federal Emergency Management Agency (FEMA) Regions and Their Member States and Territories.

Region I: Connecticut, Maine, Massachusetts, New Hampshire, Rhode Island, Vermont

Region II: (New Jersey, New York, Puerto Rico, and the U.S. Virgin Islands

Region III: (Delaware, District of Columbia, Maryland, Pennsylvania, Virginia and W. Virginia

Region IV: Alabama, Florida, Georgia, Kentucky, Mississippi, N. Carolina, S. Carolina and Tennessee

Region V: Illinois, Indiana, Michigan, Minnesota, Ohio and Wisconsin

Region VI: Arkansas, Louisiana, New Mexico, Oklahoma and Texas

Region VII: Iowa, Kansas, Missouri and Nebraska

Region VIII: Colorado, Montana, N. Dakota, S. Dakota, Utah and Wyoming

Region IX: Arizona, California, Hawaii, Nevada, American Samoa, Guam, Commonwealth of the Northern Mariana Islands, Republic of the Marshall Islands, and Federated States of Micronesia

Region X: Alaska, Idaho, Oregon and Washington

APPENDIX XIII: COMMENTS FROM THE DEPARTMENT OF HOMELAND SECURITY

U.S. Department of Homeland Security
Washington, DC 20528

Homeland
Security

August 28, 2018

Chris P. Currie
Director, Homeland Security and Justice
U.S. Government Accountability Office
441 G Street, NW
Washington, DC 20548

Re: Management Response to Draft Report 18-472, "2017 HURRICANES AND WILDFIRES: Initial Observations on the Federal Response and Key Recovery Challenges"

Dear Mr. Currie:

Thank you for the opportunity to review and comment on the subject draft report. The U.S. Department of Homeland Security (DHS) appreciates the U.S. Government Accountability Office's (GAO) work in planning and conducting its review and issuing this report.

The Department is pleased to note GAO's recognition of the Federal Emergency Management Agency (FEMA) and its partners' preparedness and coordination efforts before, during, and after the 2017 hurricanes and wildfires. Of particular note, the draft report highlighted many of the challenges that complicated response and recovery efforts. Overall, the report provides many insights to the emergency management community that stakeholders and others may not have had otherwise.

Last year's disasters were a devastating experience for millions of Americans, with more disaster survivors registering for FEMA assistance than the previous 10 years combined. The Nation responded to three major hurricanes (Harvey, Irma, and Maria) and then California wildfires in quick succession. These events came at a time when FEMA was already supporting 692 federally declared disasters and tested the Nation's ability to respond to and recover from multiple concurrent disasters. FEMA Urban Search and Rescue Task Forces, comprised of state and local emergency responders, saved or assisted in saving nearly 9,500 lives across the three hurricanes. These numbers stand in addition to the thousands of lives saved or assisted by the Department of Defense, the U.S. Coast Guard, state and local first responders, and neighbors helping neighbors. Concurrent with response operations, FEMA moved quickly to meet long-term survivor needs.

In addition to conducting concurrent response operations for Hurricanes Harvey, Irma, and Maria, FEMA took action on Hurricanes Jose and Nate. The impacts from those two storms were minimal by comparison, but nevertheless required FEMA's focus and resources. Hurricane Jose also complicated Hurricane Irma and Hurricane Maria response efforts in the Caribbean by limiting sea transport of food and water as well as transit of U.S. Naval response assets to the U.S. Virgin Islands and Puerto Rico. Hurricane Nate resulted in Major Disaster declarations in two states, Alabama and Mississippi.

The draft report also highlights the positive driving force of pre-existing relationships and coordination. During the aftermath of Hurricanes Harvey and Irma in September 2017, Acting Secretary of Homeland Security Elaine Duke activated the Surge Capacity Force for only the second time in history. This time the program was expanded beyond DHS to the entirety of the Federal Government. In total, more than 4,000 federal employees deployed and performed duties in positions within several of FEMA's incident workforce field programs. Based on this experience, FEMA has taken action to increase preparedness for the 2018 Hurricane Season by updating hurricane plans, annexes, and procedures for states, tribal lands, and territories. FEMA has also made improvements in staffing for incidents, logistics operations, and refining communications from land mobile radios to satellite communications. Finally, FEMA has updated high priority national level contracts to be better prepared to cope with responding to multiple concurrent disasters across the Nation.

FEMA is constantly reviewing its program delivery, decision-making processes, and response efforts to ensure that it can improve, minimize errors, and better serve survivors. This approach allows FEMA to build a culture of preparedness, ready the Nation for catastrophic disasters, and reduce the complexities. FEMA has reflected this commitment by incorporating objectives in its 2018-2022 Strategic Plan to include improving the readiness of incident workforce cadres; building staff, equipment, and contract capacity to achieve and maintain a higher logistics readiness rate; improving continuity and resilient communications capabilities; and streamlining the disaster survivor and grantee experience. For example, FEMA has already published alternative procedures guidance to promote outcome-based recovery in Puerto Rico by simplifying the funding process, reducing the time to receive federal funding, and providing the Commonwealth greater flexibility to rebuild its infrastructure. In addition, FEMA has initiated an update to the National Response Framework based on lessons learned in 2017 and is continuing work on a housing assistance initiative with the goal of empowering state and local governments to provide scalable disaster housing solutions tailored to the impacts of disasters and the needs of affected communities.

Disaster can strike at any time and in any place, building slowly, or occurring suddenly without warning. FEMA is part of a larger team of federal agencies, state, local, tribal, and territorial governments, and non-governmental stakeholders that share responsibility for emergency management and National preparedness. Those closest to impacted areas are the true first responders during any emergency or disaster – individuals, families, neighbors, and local communities. Whatever the scenario, FEMA serves in a coordination and integration role, collaborating with others to ensure we are a more prepared and resilient Nation.

Again, thank you for the opportunity to review and comment on the draft report. Technical comments were provided under separate cover. Please feel free to contact me if you have any questions. We look forward to working with you again in the future.

Sincerely,

JIM H. CRUMPACKER, CIA, CFE
Director
Departmental GAO-OIG Liaison Office

In: Natural Disasters
Editor: Sonja Torres

ISBN: 978-1-53616-407-7

Chapter 2

FEDERAL DISASTER ASSISTANCE AFTER HURRICANES KATRINA, RITA, WILMA, GUSTAV, AND IKE (UPDATED)*

Bruce R. Lindsay and Jared C. Nagel

ABSTRACT

This chapter provides information on federal financial assistance provided to the Gulf States after major disasters were declared in Alabama, Florida, Louisiana, Mississippi, and Texas in response to the widespread destruction that resulted from Hurricanes Katrina, Rita, and Wilma in 2005 and Hurricanes Gustav and Ike in 2008.

Though the storms happened over a decade ago, Congress has remained interested in the types and amounts of federal assistance that were provided to the Gulf Coast for several reasons. This includes how the money has been spent, what resources have been provided to the region, and whether the money has reached the intended people and entities. The financial information is also useful for congressional oversight of the federal programs provided in response to the storms. It gives Congress a general idea of the federal assets that are needed and

* This is an edited, reformatted and augmented version of Congressional Research Service, Publication No. R43139, dated February 26, 2019.

can be brought to bear when catastrophic disasters take place in the United States. Finally, the financial information from the storms can help frame the congressional debate concerning federal assistance for current and future disasters.

The financial information for the 2005 and 2008 Gulf Coast storms is provided in two sections of this chapter:

- Table 1 of Section I summarizes disaster assistance supplemental appropriations enacted into public law primarily for the needs associated with the five hurricanes, with the information categorized by federal department and agency; and
- Section II contains information on the federal assistance provided to the five Gulf Coast states through the most significant federal programs, or categories of programs.

The financial findings in this chapter include:

- Congress has appropriated roughly $121.7 billion in hurricane relief for the 2005 and 2008 hurricanes in 10 supplemental appropriations statutes.
- The appropriated funds have been distributed among 11 departments, 3 independent agencies/entities, numerous sub-entities, and the federal judiciary.
- Congress appropriated almost half of the funds ($53.8 billion, or 44% of the total) to the Department of Homeland Security, most of which went to the Disaster Relief Fund (DRF) administered by the Federal Emergency Management Agency (FEMA).
- Congress targeted roughly 22% of the total appropriations (almost $27 billion) to the Department of Housing and Urban Development for community development and housing programs.
- Approximately 20% ($25 billion) was appropriated to Department of Defense entities: $15.6 billion for civil construction and engineering activities undertaken by the Army Corps of Engineers and $9.2 billion for military personnel, operations, and construction costs.
- FEMA has reported that roughly $5.9 billion has been obligated from the DRF after Hurricanes Katrina, Rita, and Wilma to save lives and property through mission assignments made to over 50 federal entities and the American Red Cross (see Table 19), $160.4 million after Hurricane Gustav through 32 federal entities (see Table 20), and $441 million after Hurricane Ike through 30 federal entities (see Table 21). In total, federal agencies obligated roughly $6.5 billion for mission assignments after the five hurricanes.

- The Small Business Administration approved almost 177,000 applications in the region for business, home, and economic injury loans, with a total loan value of almost $12 billion (Table 31 and Table 32).
- The Department of Education obligated roughly $1.8 billion to the five states for elementary, secondary, and higher education assistance (Table 12).

This chapter also includes a brief summary of each hurricane and a discussion concerning federal to state cost-shares. Federal assistance to states is triggered when the President issues a major disaster declaration. In general, once declared the federal share for disaster recovery is 75% while the state pays for 25% of recovery costs. However, in some cases the federal share can be adjusted upward when a sufficient amount of damage has occurred, or when altered by Congress (or both). In addition, how much federal assistance is provided to states for major disasters is influenced not only by the declaration, but also by the percentage the federal government pays for the assistance. This chapter includes a cost-share discussion because some of these incidents received adjusted cost-shares in certain areas.

INTRODUCTION

This chapter provides a comprehensive summary of the federal financial assistance provided to the Gulf Coast states of Alabama, Florida, Louisiana, Mississippi, and Texas in response to the widespread destruction that resulted from Hurricanes Katrina, Rita, and Wilma in 2005 and Hurricanes Gustav and Ike in 2008.

The damages caused by the hurricanes are some of the worst in the history of the United States in terms of lives lost and property damaged and destroyed. The federal government played a significant role in the response to the hurricanes and Congress appropriated funds for a wide range of activities and efforts to help the Gulf Coast states recover and rebuild from the storms. In addition, Congress appropriated a significant amount of funds for mitigation activities and projects to reduce or eliminate the impacts of future storms.

Though the storms happened over a decade ago, Congress remains interested in the types and amounts of federal assistance that were provided to the Gulf Coast for several reasons. For one, Congress continues to be interested in how the money has been spent, what resources have been provided to the region, and whether the money has reached the people and entities intended to receive the funds. The financial information is also useful for congressional oversight and evaluation of the federal entities that were responsible for response and recovery operations. Similarly, it gives Congress a general idea of the federal assets that are needed and can be brought to bear when catastrophic disasters take place in the United States. As such, the financial information from the storms can help frame the congressional debate concerning federal assistance for current and future disasters.

The financial information provided in this chapter includes a summary of appropriations provided to the Gulf Coast states by Congress in response to the 2005 and 2008 hurricanes. In addition, when available, hurricane-specific and state-specific funding information is provided by federal entity.

BACKGROUND[1]

The 2005 hurricane season was a record-breaking season for hurricanes and storms. There were 13 hurricanes in 2005, breaking the old record of 12 hurricanes set in 1969.[2] The 2005 season also set a record for the number of category 5 storms (three) in a season.[3] Most of the damaging effects caused by the hurricanes were experienced in the Gulf Coast states of Louisiana, Arkansas, Florida, Mississippi, and Texas. The 2008 hurricane season was also an active hurricane season that caused additional damage in the Gulf Coast.

---

[1] This section was coauthored by Bruce Lindsay, Analyst in American National Government, Government and Finance Division; and Jared Nagel, Information Research Specialist, Government and Finance Division.

[2] "Colorado State U. Review Finds 2005 Hurricane Season 'Most Active,'" *Insurance Journal*, February 5, 2006.

[3] Ibid.

Hurricane Katrina

On August 23, 2005, Hurricane Katrina began about 200 miles southeast of Nassau in the Bahamas as a tropical depression. It became a tropical storm the following day. On August 24-25, 2005, the storm moved through the northwestern Bahamas and then turned westward toward southern Florida. Katrina became a hurricane just before making landfall near the Miami Dade/Broward county line during the evening of August 25, 2005. The hurricane moved southwestward across southern Florida into the eastern Gulf of Mexico on August 26, 2005. Katrina then strengthened significantly, reaching Category 5 intensity on August 28. On August 29, 2005, Hurricane Katrina made landfall in southern Plaquemines Parish, LA. The storm affected a broad geographic area—stretching from Alabama, across coastal Mississippi, to southeast Louisiana. Hurricane Katrina was reported as a category 4 storm when it initially made landfall in Louisiana, but was later downgraded to a category 3 storm. Even as a category 3 storm, Hurricane Katrina was one of the strongest storms to impact the U.S. Gulf Coast. The force of the storm was significant. The winds to the east of the storm's center were estimated to be nearly 125 mph (see Figure 1).[4]

The Gulf Coast has had a history of devastating hurricanes, but Hurricane Katrina was singular in many respects. Approximately 1.2 million people evacuated from the New Orleans metropolitan area.[5] While the evacuation helped to save lives, over 1,800 people died in the storm.[6] In addition, Hurricane Katrina destroyed or made uninhabitable an estimated 300,000 homes[7] and displaced over 400,000 citizens.[8] Economic

---

[4] National Oceanic and Atmospheric Administration, "Hurricanes in History." See Katrina, available at http://www.nhc.noaa.gov/outreach/history/#katrina.

[5] U.S. Congress, Senate Committee on Homeland Security and Governmental Affairs, *Hurricane Katrina: A Nation Still Unprepared*, 109th Cong., 2nd sess., S.Rept. 109-322 (Washington: GPO, 2006), p. 42.

[6] See National Oceanic and Atmospheric Administration, National Climate Data Center Billion-Dollar Weather/Climate Disaster website, available at http://www.ncdc.noaa.gov/billions/events.

[7] The White House, *The Federal Response to Hurricane Katrina: Lessons Learned*, February 23, 2006, p. 7, available at http://library.stmarytx.edu/acadlib/edocs/katrinawh.pdf.

[8] Kimberly A. Geaghan, *Forced to Move: An Analysis of Hurricane Katrina Movers*, U.S. Census Bureau, SEHSD Working Paper, Washington DC, June 2011, p. 1, available at

losses from the storm were estimated to be between $125 billion and $150 billion.[9]

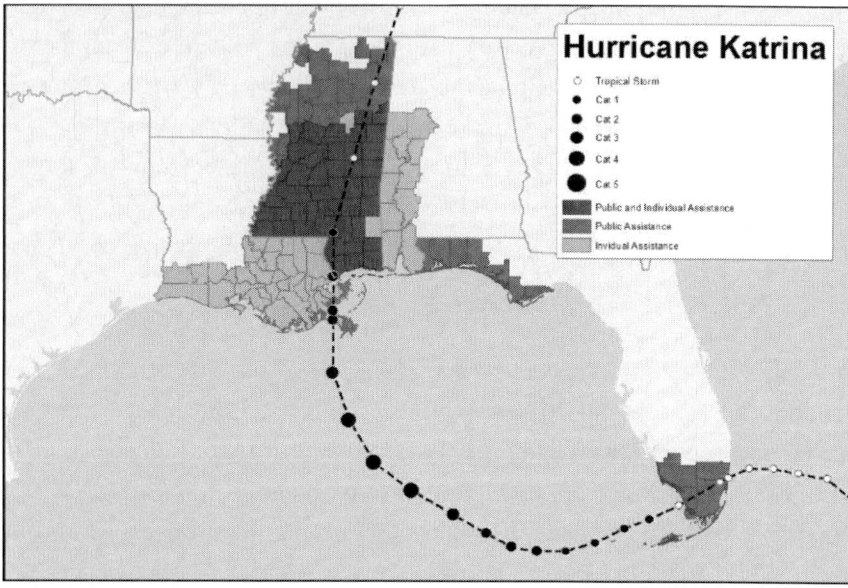

Source: Assistance information obtained from Federal Emergency Management Agency, *Disasters*, available at https://www.fema.gov/disasters; storm path information obtained from National Oceanic Atmospheric Administration, *Historical Hurricane Tracks,* available at https://coast.noaa.gov/hurricanes/.

Notes: Public Assistance covers the repairs or replacement of infrastructure (roads, bridges, public buildings, etc.) but also includes debris removal and emergency protective measures which may cover additional costs for local public safety groups incurred by their actions in responding to the disaster. Individual Assistance includes various forms of help for families and individuals following a disaster event. The assistance authorized by the Stafford Act can include housing assistance, disaster unemployment assistance, crisis counseling, and other programs intended to address the needs of people.

Figure 1. Hurricane Katrina.

---

https://www.census.gov/content/dam/Census/programssurveys/ahs/working-papers/HK_Movers-FINAL.pdf.

[9] Kristy Frame, Lynne Montgomery, and Christopher Newbury, *Bank Performance after Natural Disasters: a Historical Perspective*, Federal Deposit Insurance Corporation, January 16, 2006, available at http://www.fdic.gov/ bank/analytical/regional/ro20054q/na/2005_winter01.html.

Hurricanes Rita and Wilma

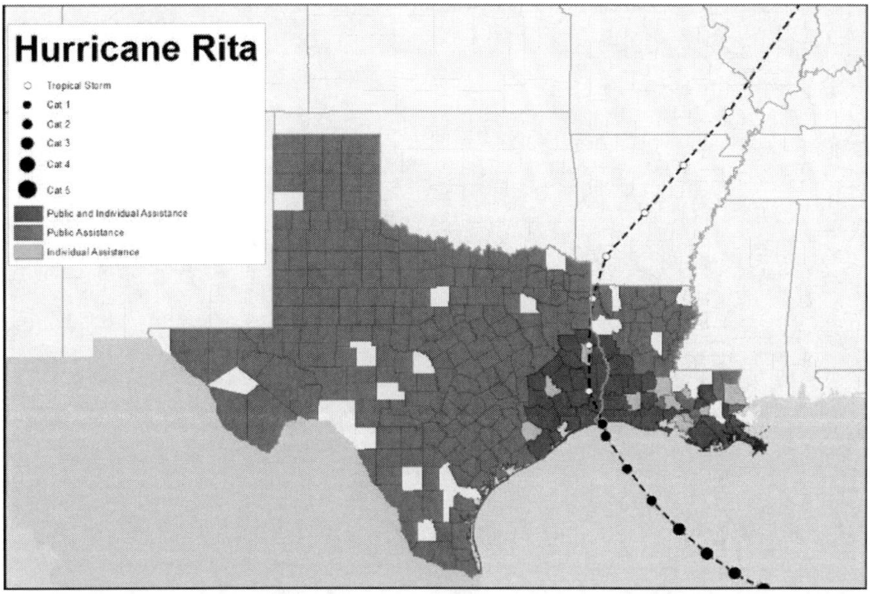

Source: Assistance information obtained from Federal Emergency Management Agency, *Disasters*, available at https://www.fema.gov/disasters; storm path information obtained from National Oceanic Atmospheric Administration, *Historical Hurricane Tracks,* available at https://coast.noaa.gov/hurricanes/.

Notes: Public Assistance covers the repairs or replacement of infrastructure (roads, bridges, public buildings, etc.) but also includes debris removal and emergency protective measures which may cover additional costs for local public safety groups incurred by their actions in responding to the disaster. Individual Assistance includes various forms of help for families and individuals following a disaster event. The assistance authorized by the Stafford Act can include housing assistance, disaster unemployment assistance, crisis counseling, and other programs intended to address the needs of people.

Figure 2. Hurricane Rita.

Two other hurricanes made landfall in the Gulf Coast shortly after Hurricane Katrina that added to recovery costs and impeded recovery efforts. On September 24, 2005, Hurricane Rita made landfall on the Texas and Louisiana border as a category 3 storm. Rita also hit parts of Arkansas and Florida. Hurricane Rita caused widespread property damage to the Gulf Coast; however, there were few deaths or injuries reported.[10]

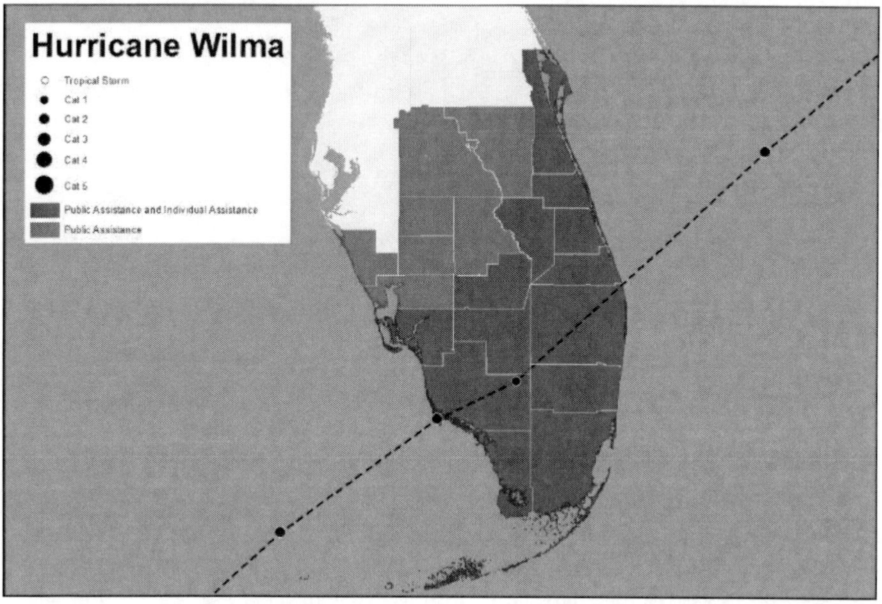

Source: Assistance information obtained from Federal Emergency Management Agency, *Disasters*, available at https://www.fema.gov/disasters; storm path information obtained from National Oceanic Atmospheric Administration, *Historical Hurricane Tracks,* available at https://coast.noaa.gov/hurricanes/.

Notes: Public Assistance covers the repairs or replacement of infrastructure (roads, bridges, public buildings, etc.) but also includes debris removal and emergency protective measures which may cover additional costs for local public safety groups incurred by their actions in responding to the disaster. Individual Assistance includes various forms of help for families and individuals following a disaster event. The assistance authorized by the Stafford Act can include housing assistance, disaster unemployment assistance, crisis counseling, and other programs intended to address the needs of people.

Figure 3. Hurricane Wilma.

---

[10] This may have been the result of Texas and Louisiana officials evacuating over 3 million residents before Rita made landfall. See National Oceanic and Atmospheric Administration, "Hurricanes in History" available at http://www.nhc.noaa.gov/outreach/history/#rita.

Rita produced rainfalls of 5 to 9 inches over large portions of Louisiana, Mississippi, and eastern Texas, with isolated amounts of 10 to 15 inches.[11] In addition, storm surge flooding and wind damage occurred in southwestern Louisiana and southeastern Texas, with some surge damage occurring in the Florida Keys (see Figure 2).[12]

On October 24, 2005, Hurricane Wilma made landfall as a Category 3 hurricane in Cape Romano, FL. The eye of Hurricane Wilma crossed the Florida Peninsula and then moved into the Atlantic Ocean north of Palm Beach (see Figure 3).[13] Hurricane Wilma killed five people in Florida and caused widespread property damage in the Gulf Coast region.

Hurricanes Gustav and Ike

In 2008, the Gulf Coast was once again affected by storms that caused billions of dollars in additional damage. On September 1, 2008, Hurricane Gustav made landfall near Cocodrie, LA, as a category 2 storm, then swept across the region causing damages in Alabama, Florida, Mississippi, and Texas (see Figure 4). Gustav produced rains over Louisiana and Arkansas that caused moderate flooding along many rivers, and is known to have produced 41 tornadoes: 21 in Mississippi, 11 in Louisiana, 6 in Florida, 2 in Arkansas, and 1 in Alabama.[14]

Hurricane Ike made landfall as a category 2 storm near Galveston, Texas, on September 13, 2008, with maximum sustained winds of 110 mph. The hurricane weakened as it moved inland across eastern Texas and Arkansas.

---

[11] National Oceanic and Atmospheric Administration, National Hurricane Center, "Hurricanes in History," available at http://www.nhc.noaa.gov/outreach/history.

[12] Ibid.

[13] Ibid.

[14] John L. Beven II and Todd B. Kimberlain, *Tropical Cyclone Report: Hurricane Gustav*, National Hurricane Center, AL072008, January 22, 2009, p. a, available at http://www.nhc.noaa.gov/data/tcr/AL072008_Gustav.pdf.

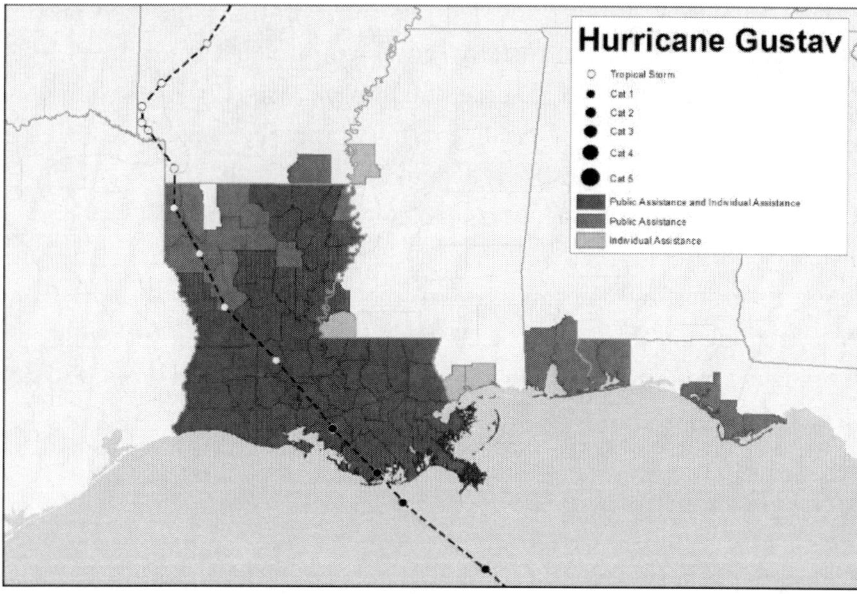

Source: Assistance information obtained from Federal Emergency Management Agency, *Disasters*, available at https://www.fema.gov/disasters; storm path information obtained from National Oceanic Atmospheric Administration, *Historical Hurricane Tracks,* available at https://coast.noaa.gov/hurricanes/.

Notes: Public Assistance covers the repairs or replacement of infrastructure (roads, bridges, public buildings, etc.) but also includes debris removal and emergency protective measures which may cover additional costs for local public safety groups incurred by their actions in responding to the disaster. Individual Assistance includes various forms of help for families and individuals following a disaster event. The assistance authorized by the Stafford Act can include housing assistance, disaster unemployment assistance, crisis counseling, and other programs intended to address the needs of people.

Figure 4. Hurricane Gustav.

Hurricane Ike's storm surge devastated the Bolivar Peninsula of Texas, and surge, winds, and flooding from heavy rains caused widespread damage in other portions of southeastern Texas, western Louisiana, and Arkansas and killed 20 people in these areas (see Figure 5).[15]

---

[15] National Oceanic and Atmospheric Administration, National Hurricane Center, "Hurricanes in History," available at http://www.nhc.noaa.gov/outreach/history.

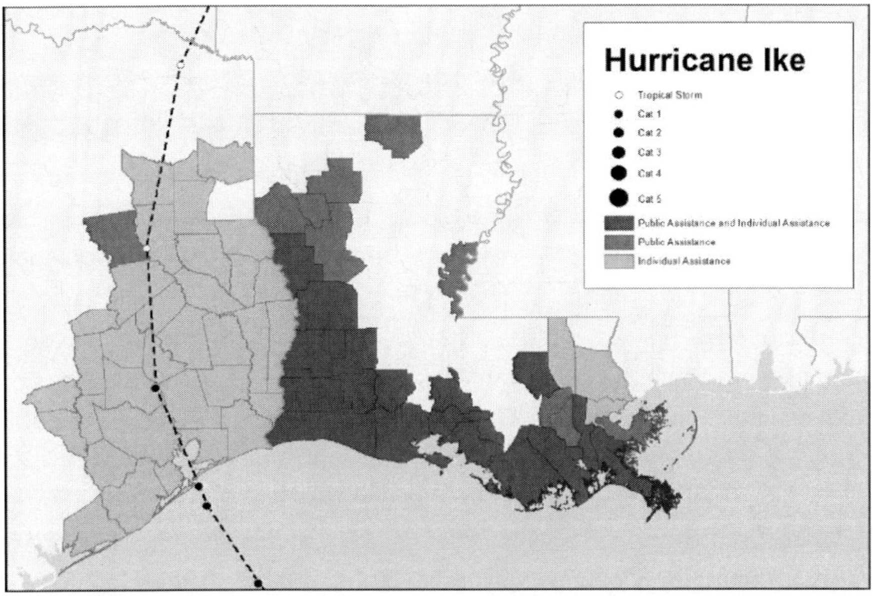

Source: Assistance information obtained from Federal Emergency Management Agency, *Disasters*, available at https://www.fema.gov/disasters; storm path information obtained from National Oceanic Atmospheric Administration, *Historical Hurricane Tracks,* available at https://coast.noaa.gov/hurricanes/.

Notes: Public Assistance covers the repairs or replacement of infrastructure (roads, bridges, public buildings, etc.) but also includes debris removal and emergency protective measures which may cover additional costs for local public safety groups incurred by their actions in responding to the disaster. Individual Assistance includes various forms of help for families and individuals following a disaster event. The assistance authorized by the Stafford Act can include housing assistance, disaster unemployment assistance, crisis counseling, and other programs intended to address the needs of people.

Figure 5. Hurricane Ike.

Additionally, as an extratropical system over the Ohio Valley, Ike was directly or indirectly responsible for 28 deaths and more than $1 billion in property damage in areas outside of the Gulf Coast.[16]

[16] Ibid.

INFORMATION CATEGORIES AND DATA COLLECTION METHODS

The following two sections provide funding data and narratives describing the assistance that was provided to the Gulf Coast in response to the 2005 and 2008 hurricane seasons. Section I presents funding provided to the five Gulf Coast states (Alabama, Florida, Louisiana, Mississippi, and Texas) after Hurricanes Katrina, Rita, Wilma, Gustav, and Ike. Funding amounts were compiled by CRS analysts who reviewed legislative texts of supplemental appropriations. The amounts are disaggregated by federal entity and sub-entity, insofar as possible and applicable. The data are based on the analysts' interpretations of disaster assistance. Some data were excluded from Section I because CRS analysts found that the data either were too ambiguous or covered disasters not limited to the Gulf Coast. Certain amounts pertaining to a range of disasters were included, however, because CRS analysts determined that most of the funds went to the Gulf Coast states.

Section II presents funding by federal agency. The amounts reported may reflect expenditures, obligations, allocations, or appropriations. The data in this section are not based solely on those in Section I. Rather, the data in Section II were derived from a variety of authoritative sources, including agency websites, CRS experts who received information directly from agencies, and governmental reports. Section II presents funding information by federal entity and includes a narrative summarizing each agency's disaster assistance efforts. The sections also provide the authorities that authorized the activities that were provided. When possible, funding data are provided in tabular form.

It should be noted that the data on appropriations in Section I, Table 1, are not directly comparable to funding data in Section II. The former were drawn solely from the public laws cited in the source note to Table 1. The data in Section II were obtained, as cited in each subsection, from a range of published and unpublished sources, and include various fiscal years.

Caveats and Limitations

Funding data on federal (and non-federal) assistance are not systematically collected. Given the absence of comprehensive federal information on disaster assistance, the data provided in this chapter should only be considered as an approximation, and should not be viewed as definitive.

In addition to the above, the following caveats apply to this chapter:

- It is difficult to identify all of the federal entities that provide disaster relief because many federal entities provide aid through a wide range of programs, not necessarily through those designated specifically as "disaster assistance" programs.[17]
- Because data on federal (and non-federal) assistance are not systematically collected, funding data were drawn from a wide-range of sources including published and unpublished data that have been collected at different times and under inconsistent reporting methods.
- Following the exodus of thousands of residents from the Gulf Coast states after Hurricane Katrina in 2005, many other states received federal assistance to cope with the influx of those seeking aid. The aid provided to the states outside the Gulf Coast is not discussed in this chapter.
- The appropriations language reviewed for Section I usually designates funds to a federal entity for a range of disasters without identifying how much funding is to be disbursed to each incident. For example, P.L. 110-329, signed into law on September 30, 2008, provided funds for several disasters that occurred in 2008, including Hurricanes Gustav and Ike, wildfires in California, and the Midwest floods. Determining the funding amounts directed toward each individual disaster is difficult, if not impossible,

---

[17] For example, the Department of Housing and Urban Development's Section 8 Housing Choice Voucher Program also provides vouchers for disaster victims.

unless the legislative text specifies these amounts. An additional difficulty occurs in tracking funding at the agency level because appropriations might be made, not to specific entities, but to budget accounts, and then allocated for specified purposes.

- The degree of transparency in reporting funding levels for disaster relief varies tremendously among federal entities. As an example, Congress requires the Federal Emergency Management Agency (FEMA) to submit monthly status reports on the Disaster Relief Fund (DRF).[18] The DRF is FEMA's disaster assistance account. The DRF is used to fund existing recovery projects (including reimbursements to other federal agencies for their work) and provide funding for future emergencies and disasters as needed. The DRF reports must detail obligations, allocations, and expenditures for Hurricanes Katrina, Rita, and Wilma. This requirement has not been extended to other agencies, and scant data exist, particularly on a state-by-state basis, on other federal funding for emergencies and major disasters.
- Appropriations may be subject to transfers or rescissions after enactment of appropriations statutes. It is possible that such emendations to the initial appropriations have not been identified in this research.

In addition to the above caveats, it should also be noted that there may have been funding changes since this chapter was originally published in 2013 that are not represented in this updated version. In some cases, additional obligations may have been provided and in other cases some funding may have been recouped or otherwise transferred. The funding information in this chapter should therefore be interpreted as illustrative as opposed to definitive, and used with appropriate caution.

[18] P.L. 110-161, codified at 42 U.S.C. §5208.

SECTION I: SUMMARY OF GULF COAST DISASTER SUPPLEMENTAL APPROPRIATIONS

Table 1. Estimated Gulf Coast Supplemental Appropriations for Hurricanes Katrina, Rita, Wilma, Gustav, and Ike (Disaster-Related Supplemental Appropriations by Department/Agency; Nominal Dollars in Millions)

Department/Agency/Program	Estimated Appropriation
Department of Agriculture	
Emergency Forestry Conservation Reserve Program	$82
Agricultural Research Service	$39
Emergency Conservation Program	$73
Farm Service Agency	$242
Executive Operations	$60
Food and Nutrition Service Commodity Assistance	$10
Forest Service	$77
Inspector General	*
Natural Resources Conservation Service	$351
Other Emergency Appropriations	*
Rural Housing Service	$90
Rural Utility Service	$53
Subtotal	$1,077
Department Of Commerce	
Department of Commerce (non specified)	$400
National Oceanic and Atmospheric Administration	$85
Marine Fishery Emergency Assistance Program	$260
Subtotal	$745
Department of Defense (Military)	
Military Personnel	$540
Operations and Maintenance	$3,684
Procurement	$2,850
Research, Development, Test, and Evaluation	$54
Military Construction and Family Housing	$1,785
Management Funds	$66
Other Defense	$236
Subtotal	$9,215a
Department of Defense (Civil)	
Army Corps of Engineers Construction	$4,951
Flood Control and Coastal Emergencies	$9,926
Flood Damage Construction for FEMA	*
Mississippi River and Tributaries	$154

Table 1. (Continued)

Department/Agency/Program	Estimated Appropriation
General Expenses	$3
Investigations	$43
Operations and Maintenance	$516
Subtotal	$15,593
Department of Education	
Office of Elementary and Secondary Education	$1,689
Office of Postsecondary Education	$292
Subtotal	$1,981
Department of Health and Human Services	
Health Resources and Services Administration	$4
Administration for Children and Families	$1,240
Centers for Disease Control and Prevention	$8
Centers for Medicare and Medicaid Services	$2,000
Subtotal	$3,252
Department of Homeland Security	
Department of Homeland Security (non-specified)	$9,157
Customs and Border Protection	$52
Federal Emergency Management Agency	$44,083b
Immigration and Customs Enforcement	$13
Office of Domestic Preparedness	$10
Office of Inspector General	$2
United States Coast Guard	$487
United States Secret Service	$4
Subtotal	$53,711
Department of Housing and Urban Development	
Community Development Block Grants	$26,200
Rental Assistance/Section 8 Vouchers	$555
Supportive Housing	$73
Public Housing Repair	$15
Office of Inspector General	$7
Subtotal	$26,850
Department of the Interior	
Department of the Interior	$210
Bureau of Reclamation	$9
Mineral Management Service	$31
National Park Service	$117
National Park Service Historical Preservation Fund	*
U.S. Fish and Wildlife Service	$162
U.S. Geological Survey	$16
Subtotal	$545

Department/Agency/Program	Estimated Appropriation
Department of Justice	
Bureau of Alcohol, Tobacco, Firearms, and Explosives	$20
Drug Enforcement Administration	$10
Federal Bureau of Investigation	$45
Federal Prison System	$11
Legal Activities	$18
Office of Justice Programs	$175
U.S. Marshals Service	$9
Subtotal	$288
Department Of Labor	
Job Corps	$16
Employment and Training Administration	$125
Subtotal	$149
Department of Transportation	
Department of Transportation (non-specified)	$722
Federal Aviation Administration	$41
Federal Highway Administration	$2,751
Federal Transportation Administration Grants	*
Maritime Administration	$8
Subtotal	$3,522
Department of Veterans Affairs	
Department Administration	$62
Veterans Health Administration	$198
Major Construction—Medical Facilities	$918
Subtotal	$1,178
Armed Forces Retirement Home	
Armed Forces Retirement Home	$242
Subtotal	$242
Environmental Protection Agency	
Environmental Protection Agency (non-specified)	$21
Subtotal	$21
General Services Administration	
General Services Administration (non-specified)	$75
Subtotal	$75
Small Business Administration	
Small Business Administration (non-specified)	$2,279
Disaster Loans Program Account	$441
Inspector General	$5
Subtotal	$2,725
The Judiciary	
The Federal Judiciary (non specified)	$18
Subtotal	$18

Table 1. (Continued)

Department/Agency/Program	Estimated Appropriation
National Aeronautics and Space Administration	
National Aeronautics and Space Administration (non-specified)	$385
Exploration Capabilities as a Consequence of Katrina	*
Subtotal	$385
Corporation For National and Community Service	
Corporation for National and Community Service	$10
Subtotal	$10
Grand Total	$121,671

Source: Data derived from CRS database of appropriations. Statutes include P.L. 109-61, P.L. 109-62, P.L. 109- 148, P.L. 109-171, P.L. 109-234, P.L. 110-28, P.L. 110-116, P.L. 110-252, P.L. 110-329, and P.L. 111-32. This table does not take into consideration any rescissions applied after Congress appropriated these funds.

Notes: * Signifies appropriation of less than $1 million. Cells marked as "non-specified" indicate appropriations funded to a department generally.

[a] This figure represents the amount appropriated after rescission of funds; it does not reflect that $1.5 billion of these funds expired in FY2006 or were transferred for other purposes.

[b] P.L. 109-62 (119 Stat. 1991) appropriated $50 billion for the Disaster Relief Fund. P.L. 109-148 (119 Stat. 2790) rescinded $23.4 billion of those funds.

Table 1 presents data on the appropriations enacted after Hurricanes Katrina, Rita, Wilma, Gustav, and Ike from FY2005 to FY2009, by federal entity and sub-entity, when possible and applicable. As mentioned earlier, in many cases funding for disaster relief is appropriated for multiple incidents. Therefore, Table 1 may include data on appropriations that also provided funding for non-Gulf Coast incidents. Some appropriations designated for a range of disasters were excluded, however, in an attempt to avoid artificially inflating the amount of funding directed to the Gulf Coast for hurricane relief.

Since FY2005, at least 10 appropriations bills have been enacted to address widespread destruction caused by the 2005 and 2008 Gulf Coast hurricanes. These appropriations consisted of eight emergency supplemental appropriations acts, one reconciliation act, and one continuing appropriations resolution.[19] In addition to these statutes that specifically identify the hurricanes or the Gulf Coast states, it is likely that

[19] These include P.L. 109-61, P.L. 109-62, P.L. 109-148, P.L. 109-171, P.L. 109-234, P.L. 110-28, P.L. 110-116, P.L. 110-252, P.L. 110-329, and P.L. 111-32.

regular appropriations legislation also provided assistance to the Gulf Coast. Because these statutes did not specify that they were providing such assistance, regular appropriations are not included in Table 1.

SECTION II. AGENCY-SPECIFIC INFORMATION ON GULF COAST HURRICANE FEDERAL ASSISTANCE

In the course of this research, CRS identified 11 federal departments, 4 federal agencies (or other entities), and numerous sub-entities, programs, and activities that supplied over $121.7 billion in federal assistance to the Gulf Coast states after the major hurricanes of 2005 (Katrina, Rita, and Wilma) and 2008 (Gustav and Ike). **Section II** provides information on the most significant programs, or categories of programs, through which the aid was provided. Each narrative contains a summary of activities of each federal entity providing disaster relief. When possible, the information is presented in tabular form and is disaster and state specific. Unless otherwise specified, all figures are stated in nominal dollars.

As mentioned earlier, the data in Section II may not correspond to the emergency funds appropriated by Congress for hurricane relief purposes specified in Section I. Reasons for the difference include the following:

- the tables in Section II present information from a variety of funding measures, including obligations, allocations, and expenditures;[20]
- some funds made available may have been reallocated or deobligated from other purposes; and
- money from accounts that did not terminate at the end of a fiscal year (known as no-year accounts) may have been allocated to the Gulf Coast states.

---

[20] For a discussion of funding terminology, see CRS Report 98-410, *Basic Federal Budgeting Terminology*, by Bill Heniff Jr.

Department of Agriculture[21]

Table 2. Disaster Relief Funding by the U.S. Department of Agriculture for 2005 Gulf Coast Hurricanes (Dollars in Thousands)

Department of Agriculture	Budget Authority	Obligations	Outlays
Agricultural Research Service	$39,000	$38,000	$37,000
Farm Service Agency			
Disaster payments-crop/livestock losses (excludes Section 32)	$132,300	$132,300	$132,300
Emergency Forestry Conservation Reserve Program (EFCRP)	$81,800	$81,800	$68,600
Emergency Conservation Program (ECP)	$84,700	$73,400	$44,800
Food and Nutrition Service	$10,000	$10,000	$9,000
Forest Service	$77,000	$77,000	$77,000
Office of Inspector General	$445	$445	$445
Natural Resources Conservation Service (NRCS)	$351,000	$300,000	$287,000
Rural Housing Service	$128,000	$101,000	$63,000
Rural Utilities Service	$53,000	$34,000	$14,000
Working Capital Fund	$60,000	$59,000	$59,000
Total	$1,017,245	$906,945	$792,145

Source: Budget Data Request No. 11-31 requested June 27, 2011, and provided July 20, 2011. Submission by Office of Budget and Program Analysis, U.S. Department of Agriculture, to Office of Management and Budget.

Notes: Figures are for Hurricanes Katrina, Rita, and Wilma in support of Gulf Coast recovery efforts and include disaster payments made under P.L. 109-234. Excludes disaster payments made under Section 32 (see Table 3) and disaster payments made under the Food, Conservation, and Energy Act of 2008 (P.L. 110-246, 2008 farm bill, see Table 4).

The U.S. Department of Agriculture (USDA) provides a variety of disaster assistance for hurricanes and other natural disasters. For the hurricanes covered in this chapter, the bulk of the department's funding has been disaster payments to producers who suffered production losses and funding for land rehabilitation programs for cleanup and restoration

[21] This section was authored by the following individuals in the Resources, Science, and Industry Division: Megan Stubbs, Specialist in Agricultural Conservation and Natural Resources Policy; Katie Hoover, Specialist in Natural Resources Policy; Tadlock Cowan, Analyst in Natural Resources and Rural Development; and Randy Alison Aussenberg, Specialist in Nutrition Assistance Policy, Domestic Social Policy Division.

projects, primarily under P.L. 109-234 and through other authorities.[22] The total USDA budget authority was over $1.0 billion for disaster relief following Hurricanes Katrina, Rita, and Wilma (Table 2). For these three hurricanes, USDA also paid an additional $112 million in farm disaster benefits to farmers in the Gulf States under various Farm Service Agency indemnity and grant programs, using funds allocated from USDA's "Section 32" Program (see "Farm Service Agency" section below).[23]

Hurricane-related support by individual agency for the 2005 and 2008 hurricanes is described in separate sections below. State-specific data are provided where available and are current as of the dates cited.

Agricultural Research Service

The Agricultural Research Service (ARS) is USDA's chief scientific research agency. Under P.L. 109-234, USDA received funding for cleanup and salvage efforts at the ARS facility in Poplarville, MS, and the Southern Regional Research Center in New Orleans, LA. Total budget authority was $39 million for the 2005 hurricanes provided under P.L. 109-234 and through reallocations from existing funds.

Farm Service Agency

The mission of the Farm Service Agency (FSA) is to serve farmers, ranchers, and agricultural partners through the delivery of agricultural support programs. Besides administering general farm commodity programs, FSA administers disaster payments for crop and livestock farmers who suffer losses from natural disasters. Following the 2005 hurricanes, producer benefits were provided under five new programs created by USDA for tropical fruit, citrus, sugarcane, nursery crops, fruits and vegetables, livestock death, feed losses, and dairy production and spoilage losses. These USDA-created programs were the Hurricane

---

[22] In many cases, these other authorities have been amended or repealed by Congress and are no longer valid. For a discussion of current agricultural disaster assistance programs, see CRS In Focus IF10565, *Federal Disaster Assistance for Agriculture*, by Megan Stubbs; CRS Report RS21212, *Agricultural Disaster Assistance*, by Megan Stubbs; or CRS Report R42854, *Emergency Assistance for Agricultural Land Rehabilitation*, by Megan Stubbs.

[23] For more information on Section 32, see CRS Report RL34081, *Farm and Food Support Under USDA's Section 32 Program*, coordinated by Jim Monke.

Indemnity Program (HIP), Livestock Indemnity Program (LIP), Feed Indemnity Program (FIP), and an Aquaculture Grant Program (AGP). Payments under the previously established Tree Indemnity Program (TIP) were provided to eligible owners of commercially-grown fruit trees, nut trees, bushes, and vines producing annual crops that were lost or damaged.[24]

Total outlays for 2005 hurricanes to the Gulf States under the aforementioned five programs were $132 million under P.L. 109-234 (see Table 2) and $112 million for four programs under "Section 32" (see Table 3 for Section 32 data). Section 32 is a permanent appropriation (originating from P.L. 74-320) that supports a variety of USDA activities, including disaster relief, federal child nutrition programs, and surplus commodity purchases.[25]

Following Hurricanes Gustav and Ike in 2008, payments were provided to qualifying producers under five nationwide agricultural disaster programs authorized in the Food, Conservation, and Energy Act of 2008 (P.L. 110-246, 2008 farm bill). Under the largest disaster program, Supplemental Revenue Assistance Payments Program (SURE),[26] the combined payments for Alabama, Florida, Louisiana, Mississippi, and Texas totaled $285 million in 2008 for a variety of natural disaster losses, including hurricane damage (Table 4). Payments for these states under the other four programs (three livestock-related programs and the Tree Assistance Program (TAP)) were approximately $66 million.[27]

---

[24] As previously stated, all of these programs have been amended, repealed, or are no longer valid. For a discussion of current agricultural disaster assistance programs, see CRS In Focus IF10565, *Federal Disaster Assistance for Agriculture*, by Megan Stubbs.

[25] USDA is currently limited in its authority to distribute emergency payments to farmers under "Section 32" authority (as well as with Commodity Credit Corporation (CCC) funds). In annual appropriations acts between FY2012 and FY2017, Congress has prohibited the use of appropriated funds to pay for salaries and expenses needed to operate a farm disaster program under either of these two funding sources. However, in FY2017 (Section 715 of FY2017 Agriculture Appropriations Act, P.L. 115-31), Congress amended this prohibition to allow such payments from available carryover funding up to $75 million.

[26] SURE authority expired in 2011 and was not reauthorized.

[27] The LFP, LIP, ELAP, and TAP programs were amended and reauthorized under section 1501 of the Agricultural Act of 2014 (P.L. 113-79, 2014 farm bill).

Table 3. 2005 Hurricane Disaster Relief Payments for Crops and Livestock by State (Dollars in Thousands)

	Alabama	Florida	Louisiana	Mississippi	North Carolina	Texas	Total
Hurricane Indemnity Program (HIP)	\$3,002	\$31,164	\$3,049	\$2,061	—	\$282	\$39,558
Tree Indemnity Program (TIP)	\$604	\$18,144	\$376	\$833	—	\$28	\$19,985
Feed Indemnity Program (FIP)	\$902	\$1,719	\$1,050	\$1,156	—	\$27	\$4,854
Livestock Indemnity Program (LIP)	\$265	\$709	\$19,238	\$2,148	—	\$701	\$23,061
Aquaculture Grant Program (AGP)	\$5,038	\$3,663	\$4,513	\$10,738	\$313	\$661	\$24,690
Total	\$9,811	\$55,399	\$28,226	\$16,936	\$313	\$1,699	\$112,384

Source: U.S. Department of Agriculture, Farm Service Agency, November 6, 2017.

Notes: Latest payment data available from USDA for 2005 hurricanes, as of November 6, 2017; the above programs were administered by FSA with funding allocated from USDA's "Section 32" Program.

FSA also administered two land rehabilitation disaster programs: (1) the Emergency Forestry Conservation Reserve Program (EFCRP),[28] which compensated private, nonindustrial forest landowners who experienced losses from hurricanes in calendar year 2005, for temporarily retiring their land; and (2) the Emergency Conservation Program (ECP),[29] which provides emergency funding and technical assistance for farmers and ranchers to rehabilitate farmland damaged by natural disasters.

---

28 Authorized by §107(a) of the Department of Defense, Emergency Supplemental Appropriations to Address Hurricanes in the Gulf of Mexico, and Pandemic Influenza Act, 2006 (P.L. 109-148), as amended. This program was repealed under section 2702(a) of the 2014 farm bill.

29 Authorized in Section 401 of the Agricultural Credit Act of 1978 (P.L. 95-334) as amended and codified under 16 U.S.C. §§2201-2205. For more information on ECP and other land rehabilitation programs, CRS Report R42854, *Emergency Assistance for Agricultural Land Rehabilitation*, by Megan Stubbs.

Table 4. 2008 Agricultural Disaster Relief Program Payments by State (Dollars in Thousands)

	Alabama	Florida	Louisiana	Mississippi	Texas	Total
Supplemental Revenue Assistance Payments Program (SURE)	$5,005	$12,932	$13,068	$4,993	$249,002	$285,000
Livestock Forage Program (LFP)	$9,002	$2,688	—	—	$40,182	$51,872
Livestock Indemnity Program (LIP)	$34	$64	$1,301	$91	$6,359	$7,849
Emergency Assistance for Livestock, Honeybees and Farm-Raised Fish Program (ELAP)	$81	$2,918	$776	$10	$659	$4,443
Tree Assistance Program (TAP)	—	$1,802	< $1	—	$146	$1,948
Total	$14,122	$20,404	$15,145	$5,094	$296,348	$351,112

Source: U.S. Department of Agriculture, Farm Service Agency, November 6, 2017.

Notes: Programs were authorized under the Food, Conservation, and Energy Act of 2008 (P.L. 110-246, 2008 farm bill). Payments as of November 6, 2017, and made for a variety of natural disaster losses that included more than just hurricane damage.

For the 2005 hurricanes, Congress provided $82 million in budget authority for EFCRP and $84.7 million in budget authority for ECP. Of the $84.7 million in budget authority for ECP, FSA obligated over $70 million. Previously unobligated funds from 2005 hurricane recovery efforts were reprogrammed in 2009 under P.L. 111-32 to be used for then current disasters, including hurricanes. On July 14, 2009, USDA announced $71 million in ECP funding, which included the 2005 reprogrammed funds, for repairing farmland damaged by natural disasters, including the hurricanes that occurred in 2008. Of the five hurricane-affected states, Texas received the largest allocation ($11 million) to address 2008 hurricane restoration efforts.

Food and Nutrition Service[30]

The Food and Nutrition Service (FNS) administers several programs that are crucial in hurricane relief efforts.[31] These include the Supplemental Nutrition Assistance Program (SNAP, formerly known as the Food Stamp Program (FSP)), child nutrition programs (e.g., school meals programs), and federally donated food commodities delivered through relief organizations. Existing laws authorize USDA to change eligibility and benefit rules to facilitate emergency aid. Disaster FSP benefits provided approximately $1 billion worth of support directly due to Hurricanes Katrina, Rita, Wilma, Gustav, and Ike.[32] Assistance provided by FSP (now, SNAP) and the child nutrition programs required no additional appropriations because the benefits are treated as entitlements.

Other than a small one-time increase in appropriations, in P.L. 109-148, to replenish some commodity stocks used for hurricane-relief purposes, no significant action was taken for hurricane relief or to pay for commodity distribution costs. This is because funding and federally provided food commodities were generally available without a need for a large appropriation.

Natural Resources Conservation Service

The Natural Resources Conservation Service (NRCS) assists private land owners with conserving soil, water, and other natural resources. Following natural disasters, NRCS works with FEMA, state and federal agencies, and local units of government to conduct post-disaster cleanup and restoration projects. NRCS administers the Emergency Watershed Protection (EWP) Program,[33] which assists landowners and operators in

---

[30] The section was authored by Randy Aussenberg, Specialist in Nutrition Assistance Policy.

[31] For further information on Food and Nutrition Service's disaster relief authorities and actions generally, see USDA-FNS website, available at http://www.fns.usda.gov/disasters/disaster.htm. For additional detail on federal food assistance provided for Hurricanes Katrina and Rita, CRS Report RL33102, *Federal Food Assistance in Disasters: Hurricanes Katrina and Rita*, by Joe Richardson.

[32] USDA, Food and Nutrition Service, 2008 Budget Explanatory Notes for Committee on Appropriations, p. 27g-13; USDA, Food and Nutrition Service, 2010 Budget Explanatory Notes for Committee on Appropriations, p. 26g-48.

[33] Authorized in §216 of P.L. 81-516 and §403 of the Agricultural Credit Act of 1978 (P.L. 95-334), as amended. Codified under 16 U.S.C. §2203 and 33 U.S.C. §701b-1. For more

implementing emergency recovery measures for slowing runoff and preventing erosion to relieve imminent hazards to life and property created by a natural disaster that causes a sudden impairment of a watershed. In the wake of 2005 and 2008 hurricane events, NRCS staff also assessed the demand and requirements for the disposal of animal carcasses, through authority delegated by FEMA. As of November 29, 2012, NRCS had obligated approximately $300 million for disaster relief stemming from these hurricanes. State EWP data for the 2005 and 2008 hurricanes are provided in Table 5 below.

Table 5. Disaster Relief Funding Through the Emergency Watershed Protection Program (Dollars in Thousands)

Hurricane	Alabama	Florida	Louisiana	Mississippi	Tennessee	Texas	Total
Katrina	$21,300	$7,200	$44,900	$114,200	$400	—	$188,000
Rita	—	—	$43,800	$2,400	—	$12,700	$58,900
Wilma	—	$12,840	—	—	—	—	$12,840
Gustav	$600	$600	$12,600	$600	—	—	$14,400
Ike	—	—	$12,000	—	—	$12,800	$24,800
Total	$21,900	$20,640	$113,300	$117,200	$400	$25,500	$298,940

Source: USDA, NRCS, November 29, 2012.

Forest Service[34]

The Forest Service (FS) administers programs for protecting and managing the natural resources of the National Forest System (NFS, primarily national forests and national grasslands) and for assisting states and non-industrial private forestland owners in protecting and managing the natural resources of non-federal forestlands. Through its State and Private Forestry (SPF) program, the FS provides financial and technical assistance, typically through state forestry agencies, to non-federal landowners to restore forests damaged by hurricanes (and other disasters).

information on EWP and other land rehabilitation programs, see CRS Report R42854, *Emergency Assistance for Agricultural Land Rehabilitation*, by Megan Stubbs.

34 This section was originally authored by Kelsi Bracmort, Specialist in Agricultural Conservation and Natural Resources Policy, Resources, Science, and Industry Division, and was updated by Katie Hoover, Specialist in Natural Resources Policy.

The state agencies are authorized to use such funds in numerous ways, such as assisting landowners to clear damaged trees and to plant new stands on cleared sites. While emergency and supplemental funding is sometimes enacted for natural disasters (e.g., hurricanes), the funding often is expended through ongoing, existing programs, and commonly cannot be distinguished from regular appropriations for these purposes (i.e., protecting and managing NFS lands and resources and assisting non-federal landowners in protecting and managing their forests).

Funding for the FS to conduct work after a natural disaster can be categorized generally as response efforts and recovery efforts. Response tasks are identified through the National Response Framework (NRF), administered by FEMA, which grants the FS certain responsibilities (e.g., firefighting) to coordinate during a presidential-declared emergency or major disaster.[35] The FS reports it spent approximately $77 million for Hurricanes Katrina, Rita, and Wilma, respectively, on response efforts in FS region 8 (state-level data was not available).[36] The FS estimates it spent a total of $2.5 million on response efforts for Hurricane Gustav ($1.4 million in Alabama, $0.9 million in Louisiana, $0.1 million in Mississippi, and $0.1 million in Texas).[37] The FS reports it spent a total of $2.1 million on response efforts for Hurricane Ike (all funding spent in Texas).

Although the FS does not have the authority for specific programs to grant recovery assistance to states, the FS can use its regular program authorities to assist state and private landowners broadly following a disaster. For example, after a hurricane, the FS may receive supplemental funding under the state and private forestry (SPF) programs appropriation to conduct recovery work via a SPF program. Eight existing FS programs were used to assist the states following Hurricanes Katrina, Rita, Wilma, Gustav, and Ike (see Table 6).[38]

---

[35] Additional information provided in the FEMA section of this report. Region 8 encompasses 13 states including the six states identified for this request.

[36] Email from the Forest Service, December 10, 2012.

[37] Email from the Forest Service, December 10, 2012.

[38] The FS reports that no funds were provided to Tennessee for any of the hurricanes.

The FS may also grant funding for the FSA Emergency Forest Restoration Program.[39] FS recovery funding amounts by state for the 2005 hurricanes (Katrina, Rita, and Wilma) and 2008 hurricanes (Gustav and Ike) are provided in Table 7.

Table 6. Forest Service Programs Used to Grant Assistance after Hurricanes in 2005 and 2008

Program	Purpose	Authority
Cooperative Forest Health Protection	Provides federal financial and technical assistance to states to facilitate their survey and monitoring of forest health conditions and for the protection of forests and trees on state and private lands from insects, disease causing agents, and invasive plants.	16 U.S.C. §2104
Economic Action Program	Assists communities and their leaders in improving the efficiency and marketing of natural resource-based industries and in diversifying rural community economic bases.	7 U.S.C. §§6611-6617
Emergency Forestry Conservation Reserve Program (temporary)	Provides assistance to nonindustrial private forest landowners who experienced a loss of 35% or more in merchantable timber from the 2005 hurricanes (Hurricane Katrina et al.).	P.L. 109-148 Section 107
Forest Stewardship	Improves timber production and environmental protection on nonfederal forest lands.	16 U.S.C. §2103a
Hazard Fuel Mitigation	Assists communities in reducing threats from wildfires.	16 U.S.C. §2106
State Fire Assistance	Provides technical and financial assistance to state cooperators.	16 U.S.C. §2106
Urban and Community Forestry	Expands knowledge and awareness of the value of urban trees and encourages the maintenance and expansion of urban tree cover.	16 U.S.C. §2105
Volunteer Fire Assistance	Provides federal financial, technical, and other assistance to state foresters and other appropriate officials to organize, train, and equip fire departments in rural areas and rural communities to prevent and suppress fires.	16 U.S.C. §2106

Source: Compiled by CRS.

[39] Additional information provided in the FSA section of this chapter.

Table 7. Forest Service 2005 and 2008 Hurricane Recovery Funding (Dollars in Thousands)

Hurricane Year	State	Program	FS Obligations
2005	Alabama	Forest Stewardship	$474
		Cooperative Forest Health Protection	$90
		Economic Action/Rural Development	$45
		State Fire Assistance	$369
		Urban and Community Forestry	$255
		Volunteer Fire Assistance	$50
Totals			$1,282
2005	Florida	Urban and Community Forestry	$615
	Louisiana	Urban and Community Forestry/ State Fire	$7,971
		Assistance/ Forest Stewardship/ Cooperative	
		Forest Health Protection[a]	
		Volunteer Fire Assistance	$517
Totals			$8,489
2005	Mississippi	Economic Action/Rural Development	$160
		Urban and Community Forestry/ State Fire	$11,519
		Assistance/ Forest Stewardship/ Cooperative	
		Forest Health Protection[a]	
		Volunteer Fire Assistance	$553
Totals			**$12,232**
2005	Texas	Economic Action/Rural Development	$83
		Urban and Community Forestry/ State Fire	$4,679
		Assistance/ Forest Stewardship/ Cooperative	
		Forest Health Protection[a]	
Totals			**$4,763**
2008	Texas	State Fire Assistance	$4,089
		Urban and Community Forestry (Carryover)	$50
		Totals	**$4,139**

Source: Data provided by the U.S. Forest Service, November 30, 2012, and confirmed on October 24, 2017.

Note: Numbers have been rounded.

[a.] Forest Service did not provide information on how these funds were allocated between the specified programs.

Rural Housing Service

The Rural Housing Service (RHS) provides loan and grant assistance for single-family and multifamily housing. RHS also administers the Community Facilities loan and grant program to provide assistance to communities for health facilities, fire and police stations, and other essential community facilities. Following the hurricanes, RHS provided housing relief to residents of the affected areas through payment moratoriums of six months, a three-month moratorium on initiating foreclosures under the single family guaranteed homeownership loans, loan forgiveness, loan reamortization, and refinancing. In addition, RHS provided temporary rental assistance to displaced family farm labor housing tenants. Assistance was provided for single-family homeowners (e.g., Section 502 loans), multi-family housing owners (e.g., Section 504 loans), and rental housing assistance (Section 521). Under P.L. 109-234, total budget authority for RHS programs for the 2005 hurricanes was $128 million.

The Disaster Relief and Recovery Supplemental Appropriations Act of 2008 (P.L. 110-329) provided funding for activities under the Rural Development Mission Area for relief and recovery from natural disasters (including hurricanes) during 2008. The act specifically provided $38 million for activities of the Rural Housing Service for areas affected by Hurricanes Katrina and Rita.

Rural Utilities Service

The Rural Utilities Service (RUS) is responsible for administering electric, telecommunications, and water assistance programs that help finance the infrastructure necessary to improve the quality of life and promote economic development in rural areas. Hurricane relief included grants for rebuilding, repairing, or otherwise improving water and waste disposal systems in designated disaster areas. Increased technical assistance under the Circuit Rider program was also provided to rural water districts. With the approval of lenders, RUS also suspended preauthorized debit payments for water and waste disposal loan guarantees for six months. Under permanent authority of P.L. 92-419, total budget authority for RUS programs for the 2005 hurricanes was $53 million.

Department of Commerce

National Oceanic and Atmospheric Administration[40]

The federal government may provide disaster relief to the fishing industry when there is a commercial fishery failure. A commercial fishery failure occurs when fishermen endure hardships resulting from fish population declines or other disruptions to the fishery. Two statutes, the Interjurisdictional Fisheries Act (16 U.S.C. §4107) and the Magnuson-Stevens Fishery Conservation and Management Act (16 U.S.C. §1864a and §1864), provide the authority and requirements for fishery disaster assistance. Under both statutes, a request for a fishery disaster determination is generally made by the governor of a state, or by a fishing community, although the Secretary of Commerce may also initiate a review at his or her own discretion. If the Secretary determines that a fishery disaster has occurred, Congress may appropriate funds for disaster assistance, which are administered by the Secretary. Funding is usually distributed as grants to states or regional marine fisheries commissions by the National Oceanic and Atmospheric Administration (NOAA) of the Department of Commerce.

Since 2005, Congress has appropriated almost $260 million of hurricane disaster relief to the Gulf of Mexico fishing industry (see Table 8). Of this total, $213 million was appropriated for damages and disruptions caused by Hurricanes Katrina and Rita (P.L. 109-234 and P.L. 110-28). Assistance provided for the direct needs of fishermen and related businesses, and supported related fisheries programs such as oyster bed and fishery habitat restoration, cooperative research, product marketing, fishing gear studies, and seafood testing. Many of these activities such as habitat restoration are ongoing management priorities for these fisheries. For damage caused by Hurricanes Gustav and Ike, $47 million was appropriated to restore damaged oyster reefs, remove storm debris, and rebuild fishing infrastructure in Texas and Louisiana (P.L. 110-329). In addition, $85 million was provided to NOAA for scanning, mapping, and

---

[40] This section was authored by Harold F. Upton, Analyst in Natural Resources Policy, Resources, Science, and Industry Division.

removing marine debris; repairing and reconstructing the NOAA Science Center; procuring a replacement emergency response aircraft and sensor package; and other activities (P.L. 109-234 and P.L. 110-28).

Table 8. Disaster Relief Funding for Commercial Fisheries (Obligations as of October 2017; Dollars in Thousands)

Commercial Fishery Disaster Assistance	Alabama	Florida	Louisiana	Mississippi	Texas	Total
Total	$44,633	$6,233	$134,190	$62,042	$11,375	$258,473

Source: NOAA Budget Office, personal communication, November 1, 2017. Gulf States Marine Fisheries Commission, Emergency Disaster Recovery Program, available at http://www.gsmfc.org/fdrp.php.

Notes: According to NOAA, all funds have been expended except for approximately $79,000. The total does not add to $260 million because $1,527 thousand was allocated for program administration. The table does not include funding for NOAA programs.

Economic Development Administration Economic Adjustment Assistance Program[41]

The Economic Development Administration (EDA) was created with the passage of the Public Works and Economic Development Act of 1965 (PWEDA), P.L. 89-136, (42 U.S.C. §3121, et. al) to provide assistance to communities experiencing long-term economic distress or sudden economic dislocation. Among the programs administered by EDA is the Economic Adjustment Assistance (EAA) program. The PWEDA (42 U.S.C. §3149(c)(2)) authorizes EDA to provide EAA funds for:

> disasters or emergencies, in areas with respect to which a major disaster or emergency has been declared under the Robert T. Stafford Disaster Relief and Emergency Assistance Act for post-disaster economic recovery.[42]

In addition to funding disaster-recovery efforts using Emergency Assistance Act (EAA) funds available under its regular appropriation, 42

---

[41] This section was authored by Eugene Boyd, Analyst in Federalism and Economic Development Policy, Government and Finance Division.

[42] Also cited as §209(c)(2) of P.L. 89-136.

U.S.C. §3233 authorizes the appropriation of such sums as are necessary to fund EAA disaster recovery activities authorized under 42 U.S.C. §3149(c)(2). Funds appropriated under 42 U.S.C. §3233 may be used to cover up to 100% of the cost of a project or activity authorized under 42 U.S.C. §3149(c)(2). Funds appropriated under a regular appropriations act may be used to cover only 50% of the cost of disaster recovery activities. However, the authorizing statute also grants EDA the authority to increase the federal share of a project's cost to 100%.

Disaster Assistance Grants

Presidentially declared disasters or emergencies are one of five specific qualifying events eligible for EAA funding assistance.[43] EAA grants are competitively awarded and may be used to help finance public facilities; public services (including job training and counseling) business development (including funding a revolving loan fund (RLF); planning; and technical assistance that support the creation or retention of private sector jobs. Regions submitting an application for EAA disaster assistance must demonstrate a clear connection between the proposed project and disaster recovery efforts. EAA disaster grants can cover 100% of a project's cost.

In order to qualify for assistance, the Secretary of Commerce must find that a proposed project or activity will help the area respond to a severe increase in unemployment, or economic adjustment problems resulting from severe changes in economic conditions. EAA regulations also require an area seeking such assistance to prepare or have in place a Comprehensive Economic Development Strategy (CEDS) outlining the nature and level of economic distress in the region, and proposed activities that could be undertaken to support private-sector job creation or retention efforts in the area.

---

[43] Other qualifying events eligible for EAA assistance, as outlined in 42 U.S.C. §3149, include communities whose economies have been injured by military-related reductions including base closures or realignments, defense contractor reductions in force, or Department of Energy defense related funding reduction; international trade; fishery failures; or the loss of manufacturing jobs.

Funding Narrative

Congress did not provide EAA supplemental appropriations for disaster recovery activities related to Hurricanes Katrina, Rita, or Wilma. However, EDA allocated $24.2 million from its regular appropriations in response to the hurricanes of 2005. In response to Hurricanes Gustav and Ike and other disasters occurring in 2008, Congress appropriated $400 million in EAA disaster supplemental funding when it approved P.L. 110-329. It also appropriated an additional $100 million in supplemental EAA disaster assistance without limiting it to disasters occurring in a specific year when it passed the Supplemental Appropriations Act of 2008, P.L. 110-252.

Of the $500 million appropriated for EAA disaster grants in 2008, EDA allocated, based on its 2010 annual report to Congress, the latest data available, a total of $63.8 million to 33 recipients in five of the six states identified in this chapter.

Department of Defense (Civil)[44]

Army Corps of Engineers

Civil Works Program

The U.S. Army Corps of Engineers (Corps) is a unique federal agency in the Department of Defense, with military and civilian responsibilities. Under its civil works program, the Corps plans, builds, operates, and maintains a wide range of water resources facilities, including hurricane protection and flood damage reduction projects, and performs emergency actions for flood and coastal emergencies.

Table 9 shows, for each Gulf Coast state, the direct appropriations that the Corps received for its water resources work related to the five

[44] This section was authored by Charles Stern, Specialist in Natural Resources Policy, Resources, Science, and Industry Division and updated by Nicole Carter, Specialist in Natural Resources Policy, Resources, Science, and Industry Division.

hurricanes. According to data the Corps provided to CRS, of the total $15.6 billion appropriated, more than $11.2 billion has been obligated.

Table 9. Disaster Relief Funding Appropriations for the Army Corps of Engineers (Dollars in Thousands)

Army Corps of Engineers	Alabama	Florida	Louisiana	Mississippi	Texas	Total
Civil Works Appropriations	$3,000	$57,000	$14,768,000	$558,000	$207,000	$15,593,000

Source: CRS correspondence with Army Corps of Engineers Budget Office, 2012.

Department of Defense (Military)[45]

Military Personnel

The Military Personnel accounts fund military pay and allowances, permanent change of station travel, retirement and health benefit accruals, uniforms, and other personnel costs. For the hurricane response efforts, funds have been used primarily to pay per diem to DOD personnel evacuated from affected areas, for the pay and allowances of activated Guard and Reserve personnel supporting the hurricane relief effort, and for increased housing allowances to compensate for housing rate increases in hurricane-affected areas. Military personnel funds obligated by the Alabama, Florida, Texas, Louisiana, and Mississippi National Guard are detailed in Table 10. Data on the obligation of other Military Personnel funds, by state, were not readily available.

Operations and Maintenance

The Operations and Maintenance (O&M) accounts fund training and operation costs, pay for civilians, maintenance service contracts, fuel,

[45] This section was authored by Lawrence Kapp, Specialist in Military Manpower Policy, Foreign Affairs, Defense, and Trade Division, with assistance from former CRS specialists Amy Belasco and Dan Else. Program summary information was taken from Department of Defense budget documents and H.Rept. 109-359, Conference Report to Accompany H.R. 2863, Department of Defense, Emergency Supplemental Appropriations to Address Hurricanes in the Gulf of MeXico, and Pandemic Influenza Act, 2006.

supplies, repair parts, and other expenses. For the hurricane response efforts, funds have been used primarily to repair facilities, establish alternate operating sites for displaced military organizations, repair and replace equipment, remove debris, clean up hazardous waste, repair utilities, evacuate DOD personnel from affected areas, and support the operations of activated Army and Air National Guard units. O&M funds obligated by the Alabama, Florida, Texas, Louisiana, and Mississippi National Guard are detailed in Table 10. Data on the obligation of other O&M funds, by state, were not readily available.

Procurement

The Procurement accounts generally fund the acquisition of aircraft, ships, combat vehicles, satellites, weapons, ammunition, and other capital equipment. For the hurricane response efforts, \$2.85 billion was appropriated, of which \$2.5 billion was used primarily to pay for extraordinary shipbuilding and ship repair costs, including not only damage to ships under construction and replacement of equipment and materials, but also additional overhead and labor costs resulting from schedule delays due to the hurricane damage to shipyards, primarily Avondale in New Orleans, Louisiana, and Ingalls in Pascagoula, Mississippi.[46] These funds also included \$140 million to improve the infrastructure at damaged shipyards.[47] Budget authority, obligations, and outlays for procurement, allocated by state for Alabama, Florida, Texas, Louisiana, and Mississippi are detailed in Table 10.

Research, Development, Test, and Evaluation

The Research, Development, Test, and Evaluation (RDT&E) accounts fund modernization efforts by way of basic and applied research, creation of technology-demonstration devices, developing prototypes, and other related costs. For the hurricane response efforts, funds have been used to

---

[46] H.Rept. 109-359, Conference Report to Accompany H.R. 2863, *Department of Defense, Emergency Supplemental Appropriations to Address Hurricanes in the Gulf of Mexico, and Pandemic Influenza Act, 2006*, p. 496.

[47] §2203, P.L. 109-234, provided that \$140 million was available for infrastructure improvements to Gulf Coast shipyards damaged in 2005.

replace damaged test equipment and repair damaged test facilities. Data allocating RDT&E funds by state were not readily available.

Military Construction (MILCON) and Family Housing

The MILCON accounts fund the acquisition, construction, installation, and equipment of temporary or permanent public works, military installations, facilities, and real property. The Family Housing Construction accounts fund costs associated with the construction of military family housing (including acquisition, replacement, addition, expansion, extension, and alteration), while the Family Housing O&M accounts fund expenses such as debt payment, leasing, minor construction, principal and interest charges, and insurance premiums on military family housing. For the hurricane response efforts, $1.4 billion was appropriated to finance the planning, design, and construction of military facilities and infrastructure that were damaged or destroyed by hurricane winds and water. Of this, $918 million was dedicated to military operations and training facilities, while an additional $460 million was appropriated for family housing construction and family housing O&M to rebuild destroyed, damaged, or new housing units and a housing office. Budget authority for MILCON and family housing construction allocated to the states of Alabama, Florida, Texas, Louisiana, and Mississippi is detailed in Table 10. Of the $1.4 billion appropriated, $1.2 billion could be allocated to the five specified states, while $167 million was devoted to planning and design activities not associated with specific locations.

Management Funds

This category includes the Defense Working Capital Fund, the National Defense Sealift Fund, and a commissary fund. For the hurricane response efforts, these funds have been used primarily to rebuild and repair damaged commissaries, replace commissary inventories, and cover transportation and contingency costs of the Defense Logistics Agency. Data allocating these funds by state were not readily available.

Other Department of Defense Programs

This category includes the Defense Health Program (DHP) and the Office of the Inspector General (OIG). The DHP title funds medical and dental care to current and retired members of the Armed Forces, their family members, and other eligible beneficiaries.

Table 10. Disaster Relief Funding by the Department of Defense (Military)(Dollars in Thousands)

Name of Program	Alabama	Florida	Louisiana	Mississippi	Texas	Total
Military Personnel (National Guard only)[a]	$7,192	$4,091	$126,982	$27,123	$15,974	$181,361
Operations and Maintenance (National Guard only)[b]	$1,407	$1,759	$89,538	$112,721	$12,440	$217,866
Procurement: Budget Authority[c]	$60,048	—	$770,647	$1,698,581	—	$2,529,277
Obligations	$60,007	—	$770,546	$1,698,193	—	$2,528,746
Outlays	$54,996	—	$697,584	$1,567,619	—	$2,320,199
Military Construction and Family Housing[d]	—	—	$371	$840	—	$1,378

Sources: The National Guard Personnel and O&M figures are CRS calculations based on data provided by the National Guard Bureau. Procurement figures were provided by the Navy. Military Construction and Family Housing figures are CRS calculations based on data contained in the conference committee reports that accompanied the relevant appropriations acts.

Notes:

[a] National Guard figures are expressed in terms of obligations.

[b] The obligated funds for National Guard personnel and O&M were for hurricane response purposes in the specified states from 2005-2012, but they may not correspond in all cases to the emergency funds appropriated by Congress for hurricane relief purposes specified in Table 1 of this chapter. An indeterminate amount of the funding came from regular appropriated funds.

[c] Procurement figures are expressed in terms of budget authority, obligations, and outlays; budget authority is nearly identical to obligations.

[d] Military construction figures are expressed in terms of budget authority; $167 million is not geographically specific.

[e] Table 11 does not include funding for Management Funds, DHP, or the OIG.

For the hurricane response efforts, these funds have been used primarily to pay for costs associated with displaced beneficiaries seeking care from private-sector providers rather than at military health care facilities, to pay the health care costs of activated Guard and Reserve personnel, and to replace medical supplies and equipment. Data allocating DHP funds by state were not readily available. Of the $589,000 appropriated for the OIG, $263,000 was provided to replace and repair damaged equipment in the Inspector General's office in Slidell, LA, and to cover relocation costs.

Department of Education[48]

Elementary and Secondary Education

Program Authorities[49]

Following the Gulf Coast hurricanes, funding to support elementary and secondary schools affected by Hurricane Katrina or Hurricane Rita was provided through three public laws: P.L. 109-148 ($1.4 billion), P.L. 109-234 ($235 million), and P.L. 110-28 ($30 million).

- P.L. 109-148 created two new programs: (1) Immediate Aid to Restart School Operations ($750 million) and (2) Temporary Impact Aid for Displaced Students ($645 million), which were specifically designed to address needs resulting from the hurricanes.[50] It also added $5 million to the McKinney-Vento

[48] This section was authored by Rebecca Skinner, Specialist in Education Policy, Domestic Social Policy Division.

[49] While not provided through education-related disaster relief legislation, Louisiana also received $20.9 million through the Charter School Program authorized under Title V-B-1 of the Elementary and Secondary Education Act specifically to help reopen charter schools damaged by Hurricanes Katrina and Rita, help create 10 new charter schools, and expand existing charter schools to accommodate displaced students. (For more information, see U.S. Department of Education, "Louisiana Awarded $20.9 Million No Child Left Behind Grant to Assist Damaged Charter Schools, Create New Charter Schools," press release, September 30, 2005, available at http://www2.ed.gov/news/pressreleases/2005/09/09302005.html).

[50] In addition to funding, P.L. 109-148 also provided general waiver authority for the Secretary of Education related to maintenance of effort (MOE) requirements; the use of federal funds to supplement, not supplant non-federal funds; and matching contributions for programs administered by the Secretary. It also modified hold harmless provisions for the Elementary

Homeless Assistance Act to serve homeless children and youth who had been displaced by the Gulf Coast hurricanes.

- P.L. 109-234 provided additional funding of $235 million for the Temporary Impact Aid for Displaced Students enacted under P.L. 109-148.
- P.L. 110-28 appropriated $30 million for elementary and secondary schools affected by the hurricanes through the Hurricane Educator Assistance program to assist in recruiting, retaining, and compensating staff in those schools.

Congress then appropriated an additional $15 million through P.L. 110-329 to provide support to local educational agencies (LEAs) whose enrollment of homeless students increased as a result of hurricanes, including Hurricanes Gustav and Ike, floods, or other natural disasters during 2008. Congress subsequently appropriated $12 million through P.L. 111-117 for the Gulf Coast Recovery Initiative to improve education in areas affected by Hurricanes Katrina, Rita, or Gustav. A brief description of each of these programs and the amount of funding each received is presented below. Table 11 details how much funding various states received under each of the programs.

Immediate Aid to Restart School Operations

The Immediate Aid to Restart School Operations provided support for LEAs and non-public schools in Louisiana, Mississippi, Alabama, and Texas to restart school operations, reopen schools, and re-enroll students. P.L. 109-148 provided $750 million for this program. This program is no longer authorized.

Temporary Emergency Impact Aid for Displaced Students

The Temporary Emergency Impact Aid for Displaced Students program provided federal funding to assist schools in enrolling students who had been displaced by the Gulf Coast hurricanes. Funds were made

---

and Secondary Education Act (ESEA) Title I-A Grants to Local Educational Agencies program and modified highly qualified teacher provisions contained in ESEA Title I-A.

available to LEAs and schools based on the number of displaced students that enrolled, irrespective of whether the school in which parents chose to enroll their child was a public or non-public school. P.L. 109-148 appropriated $645 million for this program. Subsequently, P.L. 109-234 appropriated an additional $235 million for this program, bringing the total program appropriation to $880 million.[51] Portions of the funds appropriated were provided to 49 states[52] and the District of Columbia based on the number of displaced students each enrolled. Louisiana, Texas, and Mississippi received the largest proportion of funds. This program is no longer authorized.

Hurricane Educator Assistance Program

The Hurricane Educator Assistance Program made federal funding available to Louisiana, Mississippi, and Alabama to use for recruiting, retaining, and compensating school staff who committed to work for at least three years in public elementary and secondary schools affected by Hurricanes Katrina or Rita. States were required to apply to receive funds, and the funds were allocated based on the number of public elementary and secondary schools that were closed for 19 days or more from August 29, 2005, through December 31, 2005. P.L. 110-28 provided $30 million for these purposes to Louisiana and Mississippi only. This program is no longer authorized.

McKinney-Vento Homeless Assistance Act

The McKinney-Vento Homeless Assistance Act provides funding to states to ensure that homeless children and youth are provided equal access to a free, appropriate public education in the same manner as provided other children and youth.[53] P.L. 109-148 appropriated $5 million for this program for LEAs serving homeless children and youth who had been displaced by Hurricane Katrina or Hurricane Rita. Eight states received

---

[51] Of the total appropriation for Temporary Emergency Impact Aid for Displaced Students, only $878 million was distributed, as the remaining funds were not needed by states under this program.

[52] Hawaii did not receive any funds through this program.

[53] 42 U.S.C. §11433.

funding under this program, with the largest grants provided to Texas and Louisiana.[54] While the McKinney-Vento Homeless Assistance Act continues to provide funding related to the education of homeless students, the provisions enacted specifically in response to the Gulf Coast hurricanes are no longer authorized.

Homeless Education Disaster Assistance[55]

P.L. 110-329 provided $15 million to LEAs whose enrollment of homeless students increased as a result of hurricanes, floods, or other natural disasters that occurred during 2008 and for which the President declared a major disaster under Title IV of the Stafford Act. ED was required to distribute the funds through the McKinney-Vento Homeless Assistance Act based on demonstrated need. These funds provided assistance to LEAs in Gulf Coast states affected by Hurricanes Gustav and Ike, as well as LEAs affected by natural disasters in other parts of the nation, such as flooding in the Midwest. The majority of the funds were provided to LEAs in Louisiana and Texas.[56] This program is no longer authorized.

Gulf Coast Recovery Initiative

P.L. 111-117 provided $12 million for competitive awards to LEAs located in counties in Louisiana, Mississippi, and Texas that were designated by FEMA as counties eligible for individual assistance as a result of damage caused by Hurricanes Katrina, Rita, or Gustav. The funds had to be used to improve education in areas affected by these hurricanes

---

[54] The eight states that received funds included Alabama, Arkansas, Florida, Georgia, Louisiana, Mississippi, Tennessee, and Texas.

[55] None of these funds were provided in response to the Gulf Coast hurricanes of 2005.

[56] While data were not available on the specific disasters experienced by the LEAs that received funding, data were available on the specific types of disasters for which institutions of higher education (IHEs) received funds under the Higher Education Disaster Relief program (P.L. 110-329), which also provided aid in response to natural disasters that occurred in 2008. According to these data, all IHEs in Louisiana that received funds were affected by Hurricane Gustav or Ike. Most IHEs in Texas that received funds were affected by Hurricane Ike. A few IHEs in Texas were affected by Hurricane Dolly, accounting for a relatively small portion of the funds allocated to IHEs in Texas. IHEs in Florida that received funding were affected by Tropical Storm Fay. LEAs in Iowa and Illinois received the remaining funds available to LEAs.

and had to be used for activities such as replacing instructional materials and equipment; paying teacher incentives; modernizing, renovating, or repairing school buildings; supporting charter school expansion; and supporting extended learning time activities. The majority of the funds were provided to LEAs in Louisiana. This program is no longer authorized.

Higher Education

Program Authorities

Appropriations to support institutions of higher education (IHEs) following the Gulf Coast hurricanes of 2005 were provided through P.L. 109-148 ($200 million), P.L. 109-234 ($50 million), and P.L. 110-28 ($30 million). P.L. 110-329 subsequently provided an additional $15 million for IHEs in areas affected by hurricanes, including Hurricanes Gustav and Ike, floods, or other natural disasters in 2008. Table 11 details the amount of funding allocated to various states under these provisions.

Hurricane Education Recovery

Of the $200 million provided under P.L. 109-148 for higher education, $95 million was specifically appropriated for the Louisiana Board of Regents, and $95 million was specifically appropriated for the Mississippi Institutes of Higher Learning for hurricane education recovery from the 2005 Gulf Coast hurricanes. Subsequently, P.L. 109-234 and P.L. 110-28 provided additional funds for hurricane education recovery under the Fund for the Improvement of Postsecondary Education (FIPSE), authorized by Title VII of the Higher Education Act, to assist IHEs adversely affected by the 2005 Gulf Coast hurricanes. Under both laws, funds were provided to help defray the expenses incurred by IHEs that were forced to close, relocate, or reduce their activities due to hurricane damage. Under P.L. 110-28, IHEs also were permitted to use these funds to make grants to students enrolled at these institutions on or after July 1, 2006. A total of $80 million was provided for IHEs affected by Hurricane Katrina or Hurricane Rita under the FIPSE for hurricane education recovery. The majority of funds appropriated for hurricane education recovery were provided to Mississippi and Louisiana. These activities are no longer authorized.

Funds to Assist IHEs Enrolling Displaced Students

The remaining $10 million appropriated under P.L. 109-148 for higher education disaster relief was provided to assist IHEs with unanticipated costs associated with the enrollment of students displaced as a result of Hurricane Katrina or Hurricane Rita. Overall, 99 IHEs in 24 states and the District of Columbia received funds related to the enrollment of displaced higher education students.[57] Louisiana and Texas received the largest state grants. This program is no longer authorized.

Higher Education Disaster Relief[58]

P.L. 110-329 provided an additional $15 million for IHEs that were located in an area affected by hurricanes, floods, and other natural disasters that occurred during 2008 and for which the President declared a major disaster under Title IV of the Stafford Act.[59] Funds provided through the Higher Education Disaster Relief program could be used to defray the expenses incurred by IHEs that were forced to close or relocate or whose operations were adversely affected by the natural disaster, and to provide grants to students who attended such IHEs for academic years beginning on or after July 1, 2008. The majority of these funds were provided to Louisiana and Texas for hurricane-related education disaster assistance related to Hurricanes Gustav and Ike.[60] This program is no longer authorized.

---

[57] The 24 states in which IHEs received funds included Alabama, Arizona, Arkansas, California, Colorado, Florida, Georgia, Illinois, Iowa, Kentucky, Louisiana, Maryland, Massachusetts, Michigan, Minnesota, Mississippi, Missouri, New Jersey, New York, Ohio, Tennessee, Texas, Utah, and Virginia.

[58] None of these funds were provided in response to the Gulf Coast hurricanes of 2005.

[59] Total obligations under this program were $15,028,360.

[60] IHEs in Arkansas, Colorado, Iowa, Illinois, Indiana, and Kentucky also received funds under this program.

Funding Summary

Table 11. Disaster Relief Funding Administered by the Department of Education Provided in Response to Hurricanes Katrina, Rita, Gustav, and Ike (Dollars in Thousands (cumulative obligations))

Department of Education	Alabama	Florida	Louisiana	Mississippi	Tennessee	Texas	Total
Elementary and secondary education							
Immediate Aid to Restart School Operations	$3,750	—	$445,604	$222,493	—	$78,153	$750,000
Emergency Impact Aid for Displaced Students[a]	$36,605	$27,214	$291,717	$100,787	$19,001	$250,890	$726,213
McKinney-Vento Homeless	$247	$196	$1,564	$687	$122	$1,687	$4,504
Education Assistance Program							
Hurricane Educator Assistance Program	—	—	$22,593	$7,407	—	—	$30,000
Homeless Education Disaster Assistance[b]	—	—	$1,171	—	—	$12,256	$13,427
Gulf Coast Recovery Initiative	—	—	$8,624	$2,638	—	$739	$12,000
Subtotal for elementary and secondary education	$40,602	$27,410	$771,273	$334,012	$19,123	$343,724	$1,536,144
Higher education							
Hurricane Education Recovery	$301	$1,507	$145,663	$117,878	—	$4,651	$270,000

Table 11. (Continued)

Department of Education	Alabama	Florida	Louisiana	Mississippi	Tennessee	Texas	Total
Funds to Assist Institutions of Higher Education Enrolling Displaced Students	$357	$34	$5,748	$327	$95	$1,750	$8,312
Higher Education Disaster Relief Program[c]	—	—	$3,524	—	—	$8,067	$11,591
Subtotal postsecondary education	$658	$1,541	$154,935	$118,206	$95	$14,468	$289,903
Total	$41,261	$28,952	$926,208	$452,217	$19,218	$358,192	$1,826,046

Source: Table prepared by CRS, December 11, 2012, based on published and unpublished data available from the U.S. Department of Education (ED).

Notes: Details may not add to totals due to rounding.

[a] Under the Emergency Impact Aid program, $1.9 million of the $880 million appropriated was not allocated to states, as the funds were not needed. Thus, the total appropriated amount is higher than the amount allocated and shown on the table.

[b] None of these funds were provided in response to the Gulf Coast hurricanes of 2005. While data were not available from ED on the specific disasters experienced by the LEAs that received funding, data were available on the specific types of disasters for which IHEs received funds under P.L. 110-329. According to these data, all IHEs in Louisiana that received funds were affected by Hurricane Gustav or Ike. Most IHEs in Texas that received funds were affected by Hurricane Ike. A few IHEs in Texas were affected by Hurricane Dolly, accounting for a relatively small portion of the funds allocated to IHEs in Texas. IHEs in Florida that received funding were affected by Tropical Storm Fay. Thus, all funds provided to LEAs in Louisiana and Texas were included in the table, while funds provided to LEAs in Florida were not included.

[c] Funds obligated to this account were in response to the 2008 hurricanes.

Following the Gulf Coast hurricanes of 2005, Congress appropriated $1.943 billion for ED to provide support to LEAs, schools, and IHEs in the Gulf Coast region and nationwide that were affected by Hurricane Katrina or Hurricane Rita.[61]

---

[61] For a more detailed discussion of federal education-related hurricane relief, see CRS Report R42881, *Education-Related Regulatory Flexibilities, Waivers, and Federal Assistance in*

Subsequently, FY2009 supplemental appropriations provided an additional $30 million for education-related disaster relief for LEAs and IHEs affected by natural disasters during the 2008 calendar year. Most recently, FY2010 appropriations provided an additional $12 million for LEAs located in specific areas affected by Hurricanes Katrina, Rita, or Gustav. Of the $1.985 billion provided for education-related disaster relief and administered by ED since the Gulf Coast hurricanes, nearly all of these funds ($1.826 billion, 92%) were provided to Alabama, Florida, Louisiana, Mississippi, Tennessee, and Texas in response to the 2005 and 2008 hurricanes. Table 11 details how much of this funding was allocated to each of these states for each of the programs discussed in this section.

Department of Health and Human Services

Administration for Children and Families[62]

Head Start

The federal Head Start program, authorized at 42 U.S.C. §9801 et seq., provides comprehensive early childhood development services to low-income children.[63] The program seeks to promote school readiness by enhancing the social and cognitive development of children through the provision of educational, health, nutritional, social, and other services. Federal Head Start funds are provided directly to local grantees (e.g., public and private nonprofit and for-profit agencies) rather than through states. Most children served in Head Start programs are three- and four-year olds, but services are authorized for children from birth through compulsory school age.

Response to Disasters and National Emergencies, by Boris Granovskiy and Alexandra Hegji.

[62] This section was authored by Karen Lynch, Specialist in Social Policy, Domestic Social Policy Division.

[63] For additional information, see CRS Report RL30952, *Head Start: Background and Funding*, by Karen E. Lynch.

In December 2005, Congress appropriated $90 million in supplemental Head Start funds for the costs of serving displaced children and the renovation of Head Start facilities affected by the Gulf Coast hurricanes of 2005.[64] The Department of Health and Human Services (HHS) Administration for Children and Families (ACF) reported awarding approximately $74 million of the total appropriation based on grantee requests; the remaining funds ($16 million) reverted to the U.S. Treasury Department.[65] The majority of the funds awarded to grantees ($72.5 million, or 98% of the $74 million) went to Head Start programs in Alabama, Florida, Louisiana, Mississippi, and Texas (see Table 12).

Social Services Block Grant

The Social Services Block Grant (SSBG), permanently authorized by 42 U.S.C. §1397 et seq., is a flexible source of funds that states use to support a wide variety of social services activities, ranging from child care to special services for the disabled.[66] States have broad discretion over the use of SSBG funds, which are typically allocated to states according to a population-based formula.

In December 2005, Congress appropriated $550 million in supplemental SSBG funds for necessary expenses related to the consequences of the Gulf Coast hurricanes of 2005.[67] ACF distributed these funds based on the number of FEMA registrants from Hurricanes Katrina, Rita, and Wilma, as well as the percent of individuals in poverty in each state. Funds were allocated to all states that took in evacuees, not just the states that were directly affected. The appropriations language expanded potential services for which these funds could be used to include "health services (including mental health services) and for repair,

---

[64] See Division B of P.L. 109-148. The appropriations language specified that costs of renovations may be covered "to the extent reimbursements from FEMA and insurance companies do not fully cover such costs."

[65] U.S. Department of Health and Human Services, Administration for Children and Families, "FY2008 Justification of Estimates for Appropriations Committees," February 2007, p. 91.

[66] For additional information, see CRS Report 94-953, *Social Services Block Grant: Background and Funding*, by Karen E. Lynch.

[67] See Division B of P.L. 109-148.

renovation, and construction of health facilities (including mental health facilities)."

In September 2008, Congress appropriated $600 million for necessary expenses resulting from major disasters occurring in 2008, including hurricanes, floods, and other natural disasters, as well as expenses resulting from Hurricanes Katrina and Rita.[68] ACF reserved a portion of these funds for states affected by major disasters of 2008 and a portion for states facing ongoing needs as a result of Hurricanes Katrina and Rita.[69] ACF distributed both sets of funds based on each state's share of FEMA registrants, as well as the overall population for each state. Like the previous supplemental, the 2008 supplemental appropriation again expanded potential services for which SSBG funds could be used, this time to include "health services (including mental health services) and for repair, renovation, and construction of health facilities (including mental health facilities), child care centers, and other social services facilities."

Combined, these two supplemental appropriations provided $1.150 billion for the SSBG. According to ACF, the bulk of these funds—$944 million, or 82% of the $1.150 billion—were allocated to Alabama, Florida, Louisiana, Mississippi, and Texas (see Table 12).[70]

Typically, SSBG funds are subject to a two-year expenditure period—meaning that funds must be spent by the end of the fiscal year subsequent to the fiscal year in which they were allotted to states.[71] However, most states had not spent all of their funds from either supplemental within the standard two-year period and, in both cases, Congress passed legislation

---

[68] See Division B of P.L. 110-329.

[69] For the purpose of allocating these funds, ACF counted major disasters occurring between January and September of 2008 for which FEMA Individual Assistance was authorized, plus Hurricanes Katrina and Rita.

[70] Of the $944 million, $519 million came from funds appropriated in P.L. 109-148, while $425 million came from funds appropriated in P.L. 110-329. Notably, allocations from the latter appropriation were developed based on needs resulting from a broader array of storms. In addition to accounting for Hurricanes Katrina, Rita, Gustav, and Ike, the formula for allocating these funds also took into account other major disasters of CY2008 that qualified for the FEMA Individual Assistance program, such as Tropical Storm Fay in Florida, Hurricane Dolly in Texas, and various other severe storms, tornados, and floods. For state-by-state allocation and expenditure data for these supplemental appropriations, see CRS Report 94-953, *Social Services Block Grant: Background and Funding*, by Karen E. Lynch.

[71] See §2002(c) of Title XX-A of the Social Security Act.

extending the spending deadline for these supplemental funds.[72] According to data from ACF, states had spent about $521 million (95%) of the 2005 $550 million supplemental before the extended deadline of September 30, 2009. ACF data indicate that states had spent about $522 million (87%) of the 2008 $600 million supplemental before the extended expenditure deadline of September 30, 2011. Unspent funds were to revert to the U.S. Treasury.

Table 12. Disaster Relief Funding for Programs at the HHS Administration for Children and Families (Cumulative Allocations as of July 2010; Dollars in Thousands)

HHS Administration for Children and Families	Alabama	Florida	Louisiana	Mississippi	Texas	Total
Head Start	$1,390	$114	$44,995	$22,212	$3,796	$72,507
Social Services Block Grant (SSBG)	$40,945	$89,194	$350,639	$156,535	$306,805	$944,117
Total	$42,335	$89,308	$395,634	$178,747	$310,601	$1,016,624

Source: CRS interpretation based on data from the HHS Administration for Children and Families (ACF). Head Start data are from ACF's *FY2008 Justification of Estimates for Appropriations Committees*. SSBG data are for combined supplemental allocations, based on data available at http://www.acf.hhs.gov/programs/ocs/resource/supplemental (for the 2005 supplemental) and http://www.acf.hhs.gov/programs/ocs/resource/grant-awards (for the 2008 supplemental).

Notes: Totals shown for the SSBG reflect a combination of supplemental funds appropriated by P.L. 109-148 in December 2005 and P.L. 110-329 in September 2008. Notably, the 2008 SSBG supplemental was appropriated for expenses resulting from major disasters occurring during 2008, as well as Hurricanes Katrina and Rita. Thus, the allocations shown in this table include some funds that were allocated for disasters other than Hurricanes Katrina, Rita, Wilma, Gustav, and Ike (e.g., Tropical Storm Fay and Hurricane Dolly).

According to the FY2009 SSBG annual report, states spent supplemental funds on 28 of the 29 SSBG service categories defined in federal regulation,[73] including education and training, counseling services,

[72] The expenditure deadline for the $550 million in supplemental SSBG funds appropriated in P.L. 109-148 was initially September 30, 2007. This deadline was extended, by P.L. 110-28, through September 30, 2009. The expenditure deadline for the $600 million in supplemental SSBG funds appropriated in P.L. 110-329 was initially September 30, 2010. This deadline was extended, by P.L. 111-285, through September 30, 2011.

[73] 45 C.F.R. §96.74(b).

and health-related services.[74] The FY2009 report indicated that most supplemental funds were spent in the "other services" category, including expenditures for certain construction and renovation costs, as well as costs related to certain health and mental health services. Notably, the FY2009 annual report only includes expenditures from the December 2005 supplemental appropriation.

Public Health and Medical Assistance[75]

DRF-Funded Mission Assignments

The Department of Health and Human Services (HHS) is the coordinating agency for Emergency Support Function 8 (ESF #8), Public Health and Medical Services, under the *National Response Framework*.[76] The Stafford Act authorizes reimbursements to HHS for many of its emergency or major disaster response activities, including (among others): deployment of operational assets (medical surge and mortuary teams, portable field hospitals, and the Strategic National Stockpile of drugs and medical supplies); disease surveillance; food and water safety activities; and workforce assistance to health departments. Reimbursements to HHS for mission assignments are presented in Table 17, Table 18, and Table 19.

DRF-Funded Crisis Counseling Program (CCP)

Pursuant to Section 416 of the Stafford Act, the President may provide assistance for the establishment of crisis counseling services in areas affected by declared major disasters. CCP, a program to provide short-term mental health screening, counseling, and referral services in presidentially declared disasters, is jointly administered by FEMA, the Substance Abuse and Mental Health Services Administration (SAMHSA) in HHS, and

---

[74] U.S. Department of Health and Human Services, Administration for Children and Families, "Social Services Block Grant Program Annual Report 2009, Chapter 5," available at http://archive.acf.hhs.gov/ programs/ocs/ssbg/reports/2009/index.html.

[75] This section was authored by Sarah A. Lister, Specialist in Public Health and Epidemiology, Domestic Social Policy Division.

[76] For more information on the National Response Framework see https://www.fema.gov/media-library/assets/ documents/117791.

affected states. Amounts provided to each state for the response to the Gulf Coast hurricanes are displayed in Table 13.

Federal Assistance for Health Care

In response to Hurricane Katrina, Congress authorized and appropriated a one-time program of up to $2.1 billion to cover full federal funding of the state match that would normally have been required under the Medicaid and State Children's Health Insurance (CHIP) programs, and the costs of uncompensated care, for eligible individuals from disaster-affected areas. Assistance was provided both to directly affected states and to certain states that hosted evacuees. Funding was also authorized "to restore access to health care in impacted communities," and was provided to stabilize the primary care workforce in three directly affected states: Alabama, Louisiana, and Mississippi.[77] Outlay amounts are presented in Table 14.[78]

Appropriations to Existing HHS Accounts

In response to the 2005 hurricanes, Congress provided, in emergency supplemental appropriations for affected areas, $4 million for communications equipment for community health centers, and $8 million for mosquito abatement in affected states.[79] The amounts obligated from this emergency supplemental funding are presented in Table 15.

---

[77] P.L. 109-171, Deficit Reduction Act (DRA), §6201, 120 Stat. 132-134, February 8, 2006; and P.L. 109-62, Second Emergency Supplemental Appropriations Act to Meet Immediate Needs Arising From the Consequences of Hurricane Katrina, 2005, 119 Stat. 1991, September 8, 2005. The $2 billion appropriated in the DRA was in addition to $100 million appropriated earlier to the National Disaster Medical System, some of which was also transferred for this purpose. See GAO, *Hurricane Katrina: Allocation and Use of $2 Billion for Medicaid and Other Health Care Needs*, GAO-07-67, February 28, 2007, and GAO, *Hurricane Katrina: CMS and HRSA Assistance to Sustain Primary Care Gains in the Greater New Orleans Area*, GAO-10-773R, June 30, 2010.

[78] Congress also provided $90 million in grants to states for high-risk pools that provide health insurance to individuals who are otherwise uninsurable. P.L. 109-171 (DRA), §6202, 120 Stat. 134. Almost all states were eligible and received awards under this program. Although it was not the primary focus, some states may have used the funds to provide insurance coverage to hurricane evacuees.

[79] P.L. 109-234, Emergency Supplemental Appropriations Act for Defense, the Global War on Terror, and Hurricane Recovery, 2006, 120 Stat. 463, June 15, 2006. A portion of funds for

Grants from Existing HHS Accounts

In some cases, funds available in existing HHS accounts were provided for hurricane relief. For example, the Centers for Medicare and Medicaid Services (CMS) Emergency Prescription Assistance Program provided up to $2 million in individual assistance for affected counties in Texas following Hurricane Ike. Also, the HHS Office of Minority Health provided $12 million in grants to minority-serving organizations following Hurricane Katrina. Third, SAMHSA Emergency Response Grants (SERG) provided funds to states for mental health and substance abuse services following Hurricane Katrina.[80] Amounts for SERG grants are presented in Table 12.

Administrative Waivers

The federal government funds a significant portion of the nation's health care costs, through the Medicare and Medicaid programs, veterans and Indian health care systems, and other activities. In response to the major hurricanes, HHS invoked numerous waiver authorities that allowed state, local, tribal, and private health care providers and facilities affected by the disasters to continue receiving federal health care services and/or reimbursements under altered conditions, such as the use of temporary facilities, the use of volunteer providers, and care provided to individuals not usually eligible.[81] Although these waivers did not provide new funds to disaster-affected areas, they prevented the loss of substantial federal revenues. Several HHS agencies also allowed states to reprogram federal grant funds, including from most of the grants administered by the Centers for Disease Control and Prevention (CDC).

---

communications equipment was provided to North Carolina, which deployed a field hospital to the Gulf Coast; funding was used to facilitate that aid.

[80] HHS, "HHS Provides Prescription Drug and Durable Medical Equipment Assistance for Uninsured Texas Victims of Hurricane Ike," press release, September 12, 2008; HHS, "HHS Awards Grants to Support Minority Health," press release, September 30, 2005; and HHS, "HHS Awards $600,000 in Emergency Mental Health Grants to Four States Devastated by Hurricane Katrina," press release, September 13, 2005, available at http://www.hhs.gov/news.

[81] These waiver authorities are described at, "Department of Health and Human Services, "1135 Waivers" available at https://www.phe.gov/Preparedness/legal/Pages/1135-waivers.aspx.

Table 13. Disaster Relief Funding for Crisis Counseling, Mental Health, and Substance Abuse Services (Allocations as of June 2010; Dollars in Thousands)

	Alabama		**Florida**		**Louisiana**		**Mississippi**		**Texas**		**Total by Program**	
Hurricane	CCP	SERG	CCP	SERG	CCP	SERG	CCP	SERG	CCP	SERG	CCP	SERG
Katrina	$3,019	$100	—	—	$100,436	$200	$41,101	$150	—	$150	$144,556	$600
Rita	—	—	—	—	$4,484	—	—	—	$2,709	—	$7,193	—
Wilma	—	—	$10,401	—	—	—	—	—	—	—	$10,401	—
Gustav	—	—	—	—	$16,476	—	—	—	—	—	$16,476	—
Ike	—	—	—	—	—	—	—	—	$8,267	—	$8,267	—
Total	$3,019	$100	$10,401	—	$121,396	$200	$41,101	$150	$10,976	$150	$186,893	$600

Source: Information for Hurricanes Katrina, Rita, and Wilma from FEMA, "Disaster Relief Fund: Monthly Status Report," (FY2010 Report to Congress), June 22, 2010, pp. 11-14; and HHS, "HHS Awards $600,000 in Emergency Mental Health Grants to Four States Devastated by Hurricane Katrina," press release, September 13, 2005, available at http://www.hhs.gov/news/. Information for Hurricanes Gustav and Ike provided by FEMA Office of External Affairs, July 14, 2010.

Notes: CCP is the Crisis Counseling Program. SERG is SAMHSA Emergency Response Grants. A hyphen indicates that no funds were provided. Although CCP allocations may have continued since June 2010, FEMA has not provided incident-specific funding information since that time. The SERG allocations as presented are final.

Table 14. Disaster Relief Funding for Health Care Costs and Infrastructure (Outlays as of December 31, 2012; Dollars in Thousands)

Source	Alabama	Florida	Louisiana	Mississippi	Texas
Health care costs	$240,300	$1,800	$741,100	$581,400	$33,100
Primary care stabilization	$38,300	-	$57,600	$92,800	-
Total	$278,600	$1,800	$998,700	$674,200	$33,100

Source: HHS Office of the Assistant Secretary for Financial Resources, February 26, 2013.

Notes: Authority and appropriations pursuant to the Deficit Reduction Act (DRA), Section 6201, limited to the Hurricane Katrina response. Amounts included $2.0 billion appropriated under DRA, and authority to transfer up to $100 million previously appropriated to the National Disaster Medical System (NDMS). For health care costs only, funding was provided to 21 additional states and the District of Columbia, which hosted evacuees.

Table 15. Disaster Relief Funding for Communications Equipment and Mosquito Abatement (Obligations as of July 2009; Dollars in Thousands)

Purpose	Alabama	Florida	Louisiana	Mississippi	Texas	Total
Communications equipment	$667	$667	$667	$667	$663	$3,331
Mosquito abatement	$798	-	$3,202	$2,871	$1,109	$7,980
Total	$1,465	$667	$3,869	$3,538	$1,772	$11,311

Source: HHS: Health Resources and Services Administration (HRSA) Office of Legislation; and CDC Washington Office, July 15, 2009.

Notes: Assistance provided for the response to Hurricane Katrina pursuant to P.L. 109-234. North Carolina also received a comparable award for communications equipment. On July 15, 2010, the HHS Office of the Assistant Secretary for Financial Resources confirmed that the amounts appropriated–$4 million for communications equipment and $8 million for mosquito abatement–had been fully obligated

Public Health Emergency Fund

The Secretary of HHS has authority to use a no-year fund[182] for public health emergencies.

However, the fund has not had a balance since the 1990s, so it was not available for the response to the 2005 and 2008 hurricanes.[83]

---

[82] No-year funds are available until they are expended.

Department of Homeland Security[84]

Federal Emergency Management

Agency Authority

The Stafford Act authorizes the President to issue major disaster or emergency declarations in response to incidents in the United States that overwhelm state and local governments.[85] Section 403(a)(1) of Stafford authorizes the President to direct federal resources to provide assistance essential to meeting immediate threats to life and property resulting from a major disaster.[86] Section 304 of the Stafford Act authorizes the reimbursement of other agencies from funds appropriated to the DRF for services or supplies furnished under the authority of the Stafford Act.[87]

Program Description

The primary mission of FEMA is to "reduce the loss of life and property and protect the Nation from all hazards, including natural disasters, acts of terrorism, and other man-made disasters, by leading and supporting the Nation in a risk-based, comprehensive emergency management system of preparedness, protection, response, recovery, and mitigation."[88]

FEMA provides assistance to states, local governments, tribal nations, individuals and families, and certain nonprofit organizations through the Disaster Relief Fund (DRF).[89] The more significant aid programs

---

[83] See "Federal Funding to Support an ESF-8 Response," in CRS Report RL33579, *The Public Health and Medical Response to Disasters: Federal Authority and Funding*, by Sarah A. Lister.

[84] This section was authored by Bruce R. Lindsay, Analyst in American National Government, Government and Finance Division.

[85] 42 U.S.C. §5121 et seq.

[86] 42 U.S.C. §5170b(a)(1).

[87] 42 U.S.C. §5147.

[88] Federal Emergency Management Agency, "About the Agency," available at https://www.fema.gov/about-agency.

[89] The DRF is the main account used to fund a wide variety of programs, grants, and other forms of emergency and disaster assistance to states, local governments, certain nonprofit entities, and families and individuals affected by disasters. In most cases, funding from the DRF is released after the President has issued a declaration pursuant to the Stafford Act. For further

authorized under the Stafford Act include the Public Assistance Program (PA);[90] and the Individual and Household Program (IHP), which includes Other Needs Assistance (ONA)[91] and Debris Removal,[92] the Hazard Mitigation Grant Program (HMGP),[93] and Essential Assistance.[94]

P.L. 112-175[95] requires the FEMA Administrator to provide a report by the fifth day of each month on the DRF which includes DRF funding summaries. The DRF report provides funding information by state for the 2005 and 2008 hurricanes. As shown in Table 16, the DRF report aggregates funding for Hurricanes Katrina, Rita, and Wilma.

Table 16. Disaster Relief Funding by the Federal Emergency Management Agency for Hurricanes Katrina, Rita, Wilma, Gustav, and Ike **(Cumulative Obligations as of February 5, 2013; Dollars in Millions)**

Hurricane	Alabama	Florida	Louisiana	Mississippi	Texas	Total
Katrina, Rita, and Wilma	$1,022	$233	$31,016	$10,058	$1,900	$44,229
Ike	$15	-	$329	-	$4,178	$4,522
Gustav	$19	$8	$1,544	$47	-	$1,618
Total	$1,056	$241	$32,889	$10,105	$6,078	$50,369

Source: Federal Emergency Management Agency, *Disaster Relief Fund: Monthly Report*, February 5, 2013.

FEMA Mission Assignments by Federal Entity

Mission assignments are directives from FEMA (on behalf of the requesting state) to other federal agencies to perform specific work in disaster operations on a reimbursable basis. The mission assignment contains information that is used by FEMA management to evaluate requests for assistance from states, other federal agencies, and internal

analysis on declaration process, see CRS Report R43784, *FEMA's Disaster Declaration Process: A Primer*, by Bruce R. Lindsay.

90 §406, 42 U.S.C. §5172.

91 §408, 42 U.S.C. §5174.

92 §407, 42 U.S.C. §5173.

93 §404, 42 U.S.C. §5170c.

94 §403, 42 U.S.C. §5170b.

95 Making continuing appropriations for fiscal year 2013, and for other purposes.

FEMA organizations.[96] Mission assignments are paid out of the DRF through funds appropriated to FEMA rather than funds appropriated directly to the respective agency. Table 17 contains a list of mission assignment funding by entity for Hurricanes Katrina, Wilma, and Rita. Table 18 contains mission assignment data for Hurricane Gustav and Table 19 contains mission assignment funding for Hurricane Ike. As shown in Tables 18, 19, and 20, mission assignment funding can be assigned directly to an agency, directly to an agency's program/activity, or both.

Table 17. Mission Assignment Funding by Agency: Hurricanes Katrina, Wilma, and Rita (Net Obligations, as of January 1, 2013)

Department/Agency	Obligations
Department of Agriculture	$2,573,496
Animal and Plant Health Inspection Service	$55,776
Food and Nutrition Service	$10,493
U.S. Forest Service	$162,523,398
Department of Commerce	$2,171,004
National Oceanic and Atmospheric Administration	$2,503,387
Department of Defense	$380,614,318
Army Corps of Engineers—Great Lakes and Ohio River Division	$2,522
Army Corps of Engineers—Mississippi Valley Division	$3,606,709,470
Army Corps of Engineers—South Atlantic Division	$234,037,021
Army Corps of Engineers—South Western Division	$208,521,382
National Geospatial Intelligence Agency	$1,005,796
Department of Energy	$209,373
Department of Health and Human Services	$74,004,453
Centers for Disease Control and Prevention	$15,101,893
Department of Homeland Security	
Customs and Border Protection	$15,487,544
Federal Law Enforcement Training Center	$459,381
Federal Protective Service	$182,228,449
National Communications System	$4,310,150
Transportation Security Administration	$351,511
U.S. Citizenship and Immigration Services	$304,257
U.S. Coast Guard	$183,542,905
U.S. Immigration and Customs Enforcement	$7,7,487,035
U.S. Secret Service	$8,800

[96] Department of Homeland Security, "Agency Information Collection Activities: Proposed Collection; Comment Request," 69 *Federal Register* 9350, February 27, 2004.

Department/Agency	Obligations
Department of Housing and Urban Development	$41,700,880
Department of Justice	$29,976,879
U.S. Parole Commission	$2,056,790
Department of Labor	$925,851
Occupational Safety and Health Administration	$4,958,193
Department of State	$18,101
Department of the Interior	$234,730
Bureau of Indian Affairs	$21,189
Bureau of Reclamation	$820,442
National Park Service	$52,921
U.S. Geological Survey	$471,065
Department of Transportation	$442,007,004
Federal Aviation Administration	$7,433
Department of the Treasury	$1,754,433
Department of Veterans Affairs	$2,931,612
Agency for International Development	$1,749,789
American Red Cross	$11,159
Corporation for National and Community Service	$1,028,304
Environmental Protection Agency	$264,062,645
Equal Employment Opportunity Commission	$354,546
General Services Administration	$56,410,169
National Aeronautics and Space Administration	$1,768,211
Department/Agency	**Obligations**
National Archives and Records Administration	$434,350
National Capital Planning Commission	$7,469
National Labor Relations Board	$215,543
Office of Personnel Management	$400,000
Office of the Chief Financial Officer	$70,100
Railroad Retirement Board	$5,419
Social Security Administration	$817,509
Tennessee Valley Authority	$9,039,858
U.S. Postal Service	$129,208
Total	$5,941,178,581

Source: Unpublished data provided by FEMA.

Notes: Mission Assignments were given to departments as well as the entities within some of the departments. The obligations data in the table reflect both department-wide and sub-department entity-specific obligations for mission assignments. Totals are not provided for each agency.

Table 18. Mission Assignments by Agency: Hurricane Gustav (Net Obligations, as of January 1, 2013)

Department/Agency	Obligations
Department of Agriculture	$45,000
Animal and Plant Health Inspection Service	$178,465
U.S. Forest Service	$2,750,000
Department of Commerce	
National Oceanic and Atmospheric Administration	$15,000
Department of Defense	
Army Corps of Engineers-Mississippi Valley Division	$105,349,225
Army Corps of Engineers-South Atlantic Division	$1,587,780
Army Corps of Engineers-South Western Division	$831,710
National Geospatial-Intelligence Agency	$62,000
Department of Energy	$120,000
Department of Health and Human Services	$17,476,000
Department of Homeland Security	
Customs and Border Protection	$857,000
Federal Communications Commission	$75,000
Federal Protective Service	$7,653,644
Information Analysis and Infrastructure Protection Program	$220,000
National Communications System	$48,426
Transportation Security Administration	$13,978
U.S. Coast Guard	$571,960
U.S. Immigration and Customs Enforcement	$82,411
Department of Housing and Urban Development	$140,000
Department of Justice	$1,281,144
Department of Labor	$10,000
Occupational Safety and Health Administration	$35,000
Department of State	$40,000
Department of the Interior	$20,000
National Park Service	$300,000
Department of Transportation	$621,904
Department of Treasury	$50,000
Department of Veterans Affairs	$10,000
Corporation for National and Community Service	$252,049
Environmental Protection Agency	$12,007,379
General Services Administration	$4,274,543
Tennessee Valley Authority	$3,448,894
Total	$161,322,222

Source: Unpublished data provided by FEMA.

Notes: Mission assignments were given to departments as well as the entities within some of the departments. The obligations data in the table reflect both department-wide and sub-department entity-specific obligations for mission assignments. Totals are not provided for each agency.

Table 19. Mission Assignments by Agency: Hurricane Ike (Net Obligations, as of January 1, 2013)

Department/Agency	Obligations
Department of Agriculture	$2,153,188
U.S. Forest Service	$18,990,000
Department of Defense	$25,030,836
Army Corps of Engineers—Mississippi Valley Division	$19,200,000
Army Corps of Engineers—South Atlantic Division	$7,926
Army Corps of Engineers—South Western Division	$243,230,000
National Geospatial-Intelligence Agency	$808,051
Department of Energy	$235,000
Department of Health and Human Services	$36,630,000
Department of Homeland Security	
Customs and Border Protection	$580,000
Federal Protective Service	$24,995,000
Information Analysis and Infrastructure Protection Program	$340,000
National Communication System	$135,000
Transportation Security Administration	$639,978
U.S. Coast Guard	$668,180
Department of Housing and Urban Development	$1,346,668
Department of Justice	$386,398
Department of Labor	
Occupational Safety and Health Administration	$30,000
Department of the Interior	$850,000
Bureau of Indian Affairs	$10,000
U.S. Geological Survey	$558,485
Department of Transportation	$115,597
Federal Aviation Administration	$250,000
Department of Treasury	$4,011
Department of Veterans Affairs	$260,000
Corporation for National and Community Service	$84,236
Environmental Protection Agency	$58,365,000
General Services Administration	$1,026,351
Tennessee Valley Authority	$4,350,768
Total	$441,280,673

Source: Unpublished data provided by FEMA.

Notes: Mission assignments were given to departments as well as the entities within some of the departments. The obligations data in the table reflect both department-wide and sub-department entity-specific obligations for mission assignments. Totals are not provided for each agency.

Department of Housing and Urban Development (HUD)[97]

Community Development Block Grants

Program Authority

The Community Development Block Grant (CDBG) program was first authorized as Title I of the Housing and Community Development Act of 1974, P.L. 93-383, (42 U.S. C. §5301, et al.).

Program Description

Funds are allocated by formula to states, Puerto Rico, and eligible (entitlement) communities to be used to fund eligible housing, neighborhood revitalization, and economic development activities. After funds are set aside for Indian tribes and insular areas 70% of each year's annual CDBG program appropriation must be allocated to CDBG entitlement communities, including metropolitan cities with populations of 50,000 persons or more, central cities of metropolitan areas, and statutorily defined urban counties. The remaining 30% of appropriated funds are allocated to states for distribution to non-entitlement communities.

Eligible activities must meet one of three national objectives. The activity must:

- principally benefit low or moderate income persons;
- aid in preventing or eliminating slums or blight; or
- address an imminent threat to the health or welfare of residents of an area, including disaster relief, mitigation, and long-term recovery activities.

In addition, a state or entitlement community grantee must certify that it will expend at least 70% of its CDBG allocation over a three-year period on eligible activities principally benefiting low- and moderate-income persons.

---

[97] This section was authored by Maggie McCarty, Specialist in Housing Policy and Eugene Boyd, Analyst in Federalism and Economic Development Policy.

In addition to allowing a state or entitlement community to fund disaster-recovery efforts under the CDBG's imminent threat national objective using CDBG regular appropriation, Congress has, at its discretion, appropriated additional supplemental CDBG funds in response to presidentially declared disasters. In addition to appropriating funds for disaster recovery activities, the statute authorizing the CDBG program grants the Department of Housing and Urban Development (HUD) the authority to waive or modify program regulations, except those relating to public notice, fair housing, civil rights, labor standards, environmental review, and the program's low-and moderate-income targeting requirement, when CDBG funds are used to respond to presidentially declared major disasters.[98]

Funds are allocated to states and communities to cover unmet needs not covered by state and local efforts, private insurers, and standard federal disaster programs administered by the Federal Emergency Management Agency, the Small Business Administration, and the Army Corps of Engineers. As a condition of funding, grantees are required to submit, for HUDs approval, a disaster recovery plan.

Funding

In the aggregate, the six states identified in Table 20 were awarded a total of $23.971 billion in CDBG disaster relief assistance to fund disaster relief activities in response to the five hurricanes identified in the table. Nearly 60% of this amount was allocated to Louisiana while Mississippi received approximately 30% of the total.

Five of the six states included in Table 20 received a total allocation of $19.672 billion in response to the Gulf Coast hurricanes of 2005. Louisiana received the largest share (75%) of this amount followed by Mississippi (28%), Texas (2.5%), Florida (1%), and Alabama (less than 1%).

[98] 42 U.S.C. §5321. For funds designated under this chapter by a recipient to address the damage in an area for which the President has declared a disaster under title IV of the Robert T. Stafford Disaster Relief and Emergency Assistance Act [42 U.S.C. §5170 et seq.], the Secretary may suspend all requirements for purposes of assistance under §5306 of this title for that area, except for those related to public notice of funding availability, nondiscrimination, fair housing, labor standards, environmental standards, and requirements that activities benefit persons of low and moderate income.

Table 20. Distribution of CDBG Disaster Recovery Funds for Selected States, by Disaster Declaration (Allocations as of Feb. 25, 2013; Dollars in Thousands)

Hurricane	Alabama	Florida	Louisiana	Mississippi	Tennessee	Texas	Total
Katrina-Rita-Wilma	$95,614	$182,970	$13,410,000	$5,481,221	—	$503,194	$19,672,999
Gustav	—	—	—	$2,281	—	—	$2,281
Ike	—	$81,063	$1,058.690	$6,283	$92,517	$3,057,919	$4,296,472
Total	$95,614	$264,033	$14,468,690	$5,489,785	$92,517	$3,561,113	$23,971,752

Source: HUD, available at http://portal.hud.gov/hudportal/HUD?src=/program_offices/comm_planning/ communitydevelopment/programs/drsi/activegrantee.

Notes: Allocations for Hurricanes Katrina, Wilma, and Rita were reported and presented as an aggregated total.

A total of $4.296 billion was awarded to five of six states included in Table 20 to support disaster recovery activities in response to Hurricane Ike. Texas accounted for 71% of the total followed by Louisiana (25%), Tennessee (2%), Florida (1.8%), and Mississippi (less than 1%).

Rental Assistance/Section 8 Vouchers

The Section 8 Housing Choice Voucher program, authorized at 42 U.S.C. §1437f(o), provides portable rent subsidies that low-income families can use to rent housing units offered by private market landlords. Families with vouchers contribute an income-based payment towards their rent (generally equal to 30% of a family's income), and the federal government, through local public housing authorities (PHAs), pays the landlord the difference between the tenant's contribution and the contract rent for the unit.

Congress provided over $555 million to HUD to provide rental assistance (in the form of Section 8 Housing Choice Vouchers) to families displaced by Hurricanes Katrina and Rita. The first $390 million of that amount was appropriated to HUD to provide temporary rental assistance vouchers to families that were previously assisted by HUD programs, but were displaced by the 2005 hurricanes.

Later, HUD was given a mission assignment by FEMA to begin providing rental assistance to all remaining households displaced by the 2005 hurricanes. HUD named this program the Disaster Housing Assistance Program (DHAP), and the cost of the DHAP was covered by FEMA's Disaster Relief Fund. Following Hurricane Ike, FEMA and HUD established another Disaster Housing Assistance Program (DHAP-Ike) for families displaced by that storm, also funded through the DRF under a mission assignment.

Following the first appropriation, and establishment of the mission assignments, Congress appropriated $85 million for HUD to fund the cost of ongoing, permanent Section 8 rental assistance vouchers for displaced families whose temporary housing assistance under DHAPKatrina was expiring. Congress later appropriated an additional $80 million to create new Section 8 rental assistance vouchers in the areas affected by Hurricanes Katrina and Rita.

Table 21 provides the total appropriations for disaster-related rental assistance vouchers. It does not provide allocations by state for all rental assistance funding because that information is not readily available and would be difficult to determine. Most of the funding for rental assistance was not allocated to the local public housing authorities (PHAs) administering the program by state. Rather, it was allocated based on where displaced families were living. For example, a PHA in Texas may have been administering a voucher on behalf of the Housing Authority of New Orleans for a family who was living in New Orleans before the storm, but relocated to Alabama after the storm.[99] The $80 million for new vouchers was allocated to housing authorities and Table 21 provides a break-down by state for those funds.

Supportive Housing

The Louisiana Recovery Corporation titled its recovery plan, which was primarily funded with emergency CDBG funding, the "Road Home" program. As shown in Table 21, Congress appropriated $73 million to HUD for allocation to Louisiana's Road Home program (Supportive Housing) to fund the creation of permanent supportive housing units for the elderly and persons with disabilities. Of that amount, $50 million was appropriated through an existing homeless assistance grant program that serves homeless persons with disabilities (called Shelter Plus Care) (authorized at 42 U.S.C. Chapter 119) and $23 million was appropriated through the Section 8 Housing Choice Voucher program.[100]

[99] The Housing Authority of New Orleans sustained so much damage as a result of the storm that they contracted with a PHA in Harris County, TX, to administer their voucher program for them.

[100] For more information about the Shelter Plus Care program, see CRS Report RL33764, *The HUD Homeless Assistance Grants: Programs Authorized by the HEARTH Act*, by Libby Perl; for more information about Section 8 vouchers, see CRS Report RL32284, *An Overview of the Section 8 Housing Programs: Housing Choice Vouchers and Project-Based Rental Assistance*, by Maggie McCarty.

Public Housing Repair

Low-rent public housing is federally subsidized housing owned and operated by local PHAs and available to low-income families. Several public housing developments, particularly in New Orleans, suffered severe damage after Hurricane Katrina. As shown in Table 21, Congress appropriated $15 million in emergency funding to HUD's public housing capital fund (authorized at 42 U.S.C. §1437g), which was allocated to PHAs to aid in the repair of severely damaged public housing in Louisiana.

Inspector General

Table 21. Disaster Relief Funding by the Department of Housing and Urban Development (Allocations; Dollars in Thousands)

Department of Housing And Urban Development	Alabama	Florida	Louisiana	Mississippi	Texas	Total
Rental	$6,109	$10,980	$16,908	$16,797	$27,706	$78,500
Assistance/Section 8						
Vouchersa						
Supportive Housing	—	—	$73,000	—	—	$73,000
Public Housing Repair	—	—	$15,000	—	—	$15,000
Inspector Generalb	N/A	N/A	N/A	N/A	N/A	$7,000
Total	$6,109	$10,980	$104,908	$16,797	$27,706	$173,500

Source: Table prepared by CRS. Figures are based on P.L. 109-148, P.L. 109-234, P.L. 110-28, P.L. 110-116, P.L. 110-252, P.L. 110-329, and P.L. 111-32. Community Development Block Grant allocations taken from https://web.archive.org/web/20120808044154/ http://portal.hud.gov:80/hudportal/HUD?src=/program_offices/comm_planning/communitydevelopment/programs/drsi/activegrantee. Rental Assistance/Section 8 Voucher allocations taken from https://web.archive.org/web/20170222002106/https:/www.hud.gov/offices/pih/publications/sec1203/thu-cong-ntf.pdf. Public Housing repair information was taken from HUD's FY2010 Congressional Budget Justifications.

Notes: Total amounts allocated do not equal total amounts appropriated in some cases because funds have been reserved by the department for administrative costs.

[a] Note that state allocations are only provided for the $80 million provided for new vouchers by P.L. 111-32.

[b] An additional $7 million provided by P.L. 109-234 for Community Development Block Grants was required to be transferred to the Office of Inspector General.

As shown in Table 21, Congress appropriated $7 million for the HUD Inspector General to help fund the cost of enhanced oversight over disaster recovery funding.

Department of Justice[101]

Established by an "Act to Establish the Department of Justice"[1102] with the Attorney General at its head, the Department of Justice (DOJ) provides counsel for the government in federal cases and protects citizens through law enforcement. It represents the federal government in all proceedings, civil and criminal, before the U.S. Supreme Court. In legal matters, generally, the department provides legal advice and opinions, upon request, to the President and executive branch department heads.

To date, the DOJ has received a total of $287.5 million in supplemental appropriations for departmental expenses related to hurricanes in the Gulf of Mexico and to award grants to Gulf Coast states. Table 22 provides a breakdown of how DOJ obligated disaster funding amongst Alabama, Florida, Louisiana, Mississippi, and Texas.

Legal Activities

Program Authority or Authorities

Subtitle A of Title XI of the Violence Against Women and Department of Justice Reauthorization Act of 2005 (P.L. 109-162) authorized appropriations for the General Legal Activities and U.S. Attorneys accounts. For the General Legal Activities account the act authorized $679.7 million for FY2006, $706.8 million for FY2007, $735.1 million for FY2008, and $764.5 million for FY2009. For the U.S. Attorneys account the act authorized $1.626 billion for FY2006, $1.691 billion for FY2007, $1.795 billion for FY2008, and $1.829 billion for FY2009.

---

[101] This section was authored by Nathan James, Analyst in Crime Policy, Domestic Social Policy Division.

[102] 28 U.S.C. §501.

Program Description

The Legal Activities account includes several sub-accounts, including General Legal Activities and the U.S. Attorneys. The General Legal Activities sub-account funds the Solicitor General's supervision of DOJ's conduct in proceedings before the Supreme Court. It also funds several departmental divisions (tax, criminal, civil, environment and natural resources, legal counsel, civil rights, INTERPOL, and dispute resolution). The U.S. Attorneys enforce federal laws through prosecution of criminal cases and represent the federal government in civil actions in all of the 94 federal judicial districts.[103]

Funding Narrative

Since 2005, Congress has appropriated a total of $17.5 million in supplemental appropriations for this account. This amount included $2.0 million for General Legal Activities and a total of $15.5 million for the U.S. Attorneys. Chapter 8 of Title II of the Emergency Supplemental Appropriations Act for Defense, the Global War on Terror and Hurricane Recovery, 2006 (P.L. 109-234) provided $2 million for General Legal Activities "to investigate and prosecute fraud cases related to hurricanes in the Gulf Coast region."[104] Chapter 8 of Title I of Division B of the Department of Defense, Emergency Supplemental Appropriations to Address Hurricanes in the Gulf of Mexico, and Pandemic Influenza Act, 2006 (P.L. 109-148) provided $9 million for the U.S. Attorneys "to support operational recovery from hurricane-related damage in the Gulf Coast region."[105] Chapter 8 of Title II of the Emergency Supplemental Appropriations Act for Defense, the Global War on Terror and Hurricane Recovery, 2006 (P.L. 109-234) provided the U.S. Attorneys with $6.5

---

[103] U.S. Department of Justice, Offices of the United States Attorneys, "United States Attorneys' Mission Statement," available at http://www.justice.gov/usao/about/mission.html.

[104] U.S. Congress, House, *Making Emergency Supplemental Appropriations for the Fiscal Year Ending September 30, 2006 and Other Purposes*, Conference Report, 109th Cong., 2nd sess., June 8, 2006, H.Rept. 109-494 (Washington: GPO, 2006), p. 128.

[105] U.S. Congress, House, *Making Appropriations for the Department of Defense for the Fiscal Year Ending September 30, 2006, and Other Purposes*, Conference Report, 109th Cong., 1st sess., December 18, 2005, H.Rept. 109-359 (Washington: GPO, 2005), p. 514.

million "to investigate and prosecute fraud cases related to hurricanes in the Gulf Coast region."[106]

United States Marshals Service

Program Authority or Authorities

Subtitle A of Title XI of the Violence Against Women and Department of Justice Reauthorization Act of 2005 (P.L. 109-162) authorized $800.3 million for FY2006, $832.3 million for FY2007, $865.6 million for FY2008, and $900.2 million for FY2009 for the United States Marshals Service (USMS) account.

Program Description

The federal marshals' service was founded in 1789, making it the oldest federal law enforcement agency. A presidentially appointed U.S. marshal directs the operations of the marshals' services in each of the 94 federal judicial districts. The USMS facilitates the functioning of the federal judicial process by providing protection for judges, attorneys, witnesses, and jurors and providing physical security in courthouses. The USMS is the federal government's primary agency for fugitive investigations. USMS task forces combine the efforts of federal, state, and local law enforcement agencies to locate and arrest fugitives. The Marshals Service also works with international law enforcement agencies to apprehend fugitives who have fled abroad and to apprehend foreign fugitives who have entered the United States. The USMS executes all federal arrest warrants. The USMS manages and sells assets which were seized or forfeited by federal law enforcement agencies.[107] The assets managed and sold by the USMS are assets that represent the proceeds of,

---

[106] U.S. Congress, House, *Making Emergency Supplemental Appropriations for the Fiscal Year Ending September 30, 2006 and Other Purposes*, Conference Report, 109th Cong., 2nd sess., June 8, 2006, H.Rept. 109-494 (Washington: GPO, 2006), p. 128.

[107] U.S. Department of Justice, U.S. Marshals Service, "Overview of the U.S. Marshals Service," available at https://www.usmarshals.gov/duties/factsheets/index.html.

or were used to facilitate federal crimes.[108] The Marshals Service is responsible for housing and transporting all federal detainees from the time they are arrested until they are either acquitted or convicted and delivered to their designated federal prison. The USMS operates the Justice Prisoner and Alien Transportation System (JPATS), which transports prisoners between judicial districts, correctional facilities, and foreign countries. The USMS is also responsible for administering the federal witness security program, which provides for the security and safety of government witnesses and their authorized family members, whose lives are in danger as a result of their cooperation with the U.S. government.[109]

Funding Narrative

Since 2005, Congress has appropriated $9 million in supplemental appropriations for the U.S. Marshal's Service. Chapter 8 of Title I of Division B of the Department of Defense, Emergency Supplemental Appropriations to Address Hurricanes in the Gulf of Mexico, and Pandemic Influenza Act, 2006 (P.L. 109-148) provided $9 million for the USMS's salaries and expenses account "to support operational recovery from hurricane-related damage in the Gulf Coast region."[110]

Federal Bureau of Investigation

Program Authority or Authorities

Subtitle A of Title XI of the Violence Against Women and Department of Justice Reauthorization Act of 2005 (P.L. 109-162) authorized $5.761 billion for FY2006, $5.992 billion for FY2007, $6.231 billion for FY2008, and $6.481 billion for FY2009 for the Federal Bureau of Investigation (FBI) account.

---

108 U.S. Department of Justice, "Assets Forfeiture Program," available at http://www.justice.gov/jmd/afp/.

109 U.S. Department of Justice, U.S. Marshals Service, "Overview of the U.S. Marshals Service," available at https://www.usmarshals.gov/duties/factsheets/overview.pdf.

110 U.S. Congress, House, *Making Appropriations for the Department of Defense for the Fiscal Year Ending September 30, 2006, and Other Purposes*, Conference Report, 109th Cong., 1st sess., December 18, 2005, H.Rept. 109-359 (Washington: GPO, 2005), p. 514.

Program Description

The FBI was founded in 1908. Its headquarters is in Washington, DC, and it has 56 field offices located in major cities throughout the United States and its territories and another 380 resident agencies in cities and towns across the nation. In addition, the FBI has more than 60 international offices called "legal attachés" in U.S. embassies worldwide. The FBI is the lead federal investigative agency charged with defending the country against foreign terrorist and intelligence threats; enforcing federal criminal laws; and providing leadership and criminal justice services to federal, state, municipal, tribal, and territorial law enforcement agencies and partners. The FBI focuses on protecting the United States from internal and external threats and investigations that are too large or too complex for state and local authorities to handle on their own. The priorities of the FBI include:

- protecting the United States from terrorist attack;
- protecting the United States against foreign intelligence operations and espionage;
- protecting the United States against cyber-based attacks and high-technology crimes;
- combating public corruption;
- protecting civil rights;
- investigating transnational/national criminal organizations and enterprises;
- investigating major white-collar crime;
- investigating significant violent crime; and
- supporting federal, state, local and international partners.[111]

The FBI collects and disseminates national crime data through the Uniform Crime Reports (UCR).[112] The FBI also operates several national

---

111 U.S. Department of Justice, Federal Bureau of Investigation, "About," available at https://www.fbi.gov/about/ mission.

112 U.S. Department of Justice, Federal Bureau of Investigation, "Uniform Crime Reports," available at https://ucr.fbi.gov/.

law enforcement information sharing systems such as the Combined DNA Index System (CODIS),[113] the Law Enforcement National Data Exchange (N-Dex),[114] the Next Generation Identification System (NGI),[115] the National Instant Criminal Background Check System (NICS),[116] and the National Crime Information Center (NCIC).[117]

Funding Narrative

Since 2005, Congress has appropriated $45 million in supplemental appropriations for the FBI. Chapter 8 of Title I of Division B of the Department of Defense, Emergency Supplemental Appropriations to Address Hurricanes in the Gulf of Mexico, and Pandemic Influenza Act, 2006 (P.L. 109-148) provided $45 million for the FBI's salaries and expenses account "to support operational recovery from hurricane-related damage in the Gulf Coast region."[118]

Drug Enforcement Administration

Program Authority or Authorities

Subtitle A of Title XI of the Violence Against Women and Department of Justice Reauthorization Act of 2005 (P.L. 109-162) authorized $1.716 billion for FY2006, $1.785 billion for FY2007, $1.856 billion for FY2008, and $1.930 billion for FY2009 for the Drug Enforcement Administration (DEA) account.

---

113 U.S. Department of Justice, Federal Bureau of Investigation, "Combined DNA Index System (CODIS)," available at http://www.fbi.gov/about-us/lab/biometric-analysis/codis.

114 U.S. Department of Justice, Federal Bureau of Investigation, "N-Dex: National Law Enforcement Data Exchange," available at http://www.fbi.gov/about-us/cjis/n-dex.

115 U.S. Department of Justice, Federal Bureau of Investigation, "Next Generation Identification (NGI)," available at https://www.fbi.gov/services/cjis/fingerprints-and-other-biometrics/ngi.

116 U.S. Department of Justice, Federal Bureau of Investigation, "National Instant Criminal Background Check System," available at http://www.fbi.gov/about-us/cjis/nics.

117 U.S. Department of Justice, Federal Bureau of Investigation, "National Crime Information Center," available at https://www.fbi.gov/services/cjis/ncic.

118 U.S. Congress, House, *Making Appropriations for the Department of Defense for the Fiscal Year Ending September 30, 2006, and Other Purposes*, Conference Report, 109th Cong., 1st sess., December 18, 2005, H.Rept. 109-359 (Washington: GPO, 2005), p. 514.

Program Description

The DEA was established in 1973 through an executive order issued by President Nixon.[119] The DEA has 226 domestic and 85 foreign offices.[120] The DEA's mission is "to enforce the controlled substances laws and regulations of the United States and bring to the criminal and civil justice system of the United States, or any other competent jurisdiction, those organizations and principal members of organizations, involved in the growing, manufacture, or distribution of controlled substances appearing in or destined for illicit traffic in the United States; and to recommend and support non-enforcement programs aimed at reducing the availability of illicit controlled substances on the domestic and international markets."[121] The DEA's primary responsibilities include:

- investigating major violators of controlled substance laws operating at interstate and international levels;
- management of a national drug intelligence program in cooperation with federal, state, local, and foreign officials to collect, analyze, and disseminate strategic and operational drug intelligence information;
- seizure and forfeiture of assets derived from, traceable to, or intended to be used for illicit drug trafficking;
- enforcement of the provisions of the Controlled Substances Act as they pertain to the manufacture, distribution, and dispensing of legally produced controlled substances;
- reduction of illicit drugs on the United States market through methods such as crop eradication, crop substitution, and training of foreign officials; and

---

[119] U.S. Department of Justice, Drug Enforcement Administration, "DEA History," available at https://www.dea.gov/ about/history.shtml.

[120] U.S. Department of Justice, Drug Enforcement Administration, "Domestic Office Locations," available at https://www.dea.gov/about/Domesticoffices.shtml. U.S. Department of Justice, Drug Enforcement Administration, Foreign Office Locations, available at https://www.dea.gov/about/foreignoffices.shtml.

[121] U.S. Department of Justice, Drug Enforcement Administration, "DEA Mission Statement," available at https://www.dea.gov/about/mission.shtml.

- liaison with the United Nations, Interpol, and other organizations on matters relating to international drug control programs.[122]

Funding Narrative

Since 2005, Congress has appropriated $10 million in supplemental appropriations for this account. Chapter 8 of Title I of Division B of the Department of Defense, Emergency Supplemental Appropriations to Address Hurricanes in the Gulf of Mexico, and Pandemic Influenza Act, 2006 (P.L. 109-148) provided $10 million for the DEA's salaries and expenses account "to support operational recovery from hurricane-related damage in the Gulf Coast region."[123]

Bureau of Alcohol, Tobacco, Firearms, and Explosives

Program Authority or Authorities

Subtitle A of Title XI of the Violence Against Women and Department of Justice Reauthorization Act of 2005 (P.L. 109-162) authorized $923.6 million for FY2006, $960.6 million for FY2007, $999.0 million for FY2008, and $1.039 billion for FY2009 for the Bureau of Alcohol, Tobacco, Firearms, and Explosives (ATF) account.

Program Description

The ATF enforces federal criminal law related to the manufacture, importation, and distribution of alcohol, tobacco, firearms, and explosives.[124] The ATF's responsibilities were transferred from the Department of the Treasury to the Department of Justice as a part of the Homeland Security Act (P.L. 107-296).[125] The ATF works both independently and through partnerships with industry groups, international,

---

[122] Ibid.

[123] U.S. Congress, House, *Making Appropriations for the Department of Defense for the Fiscal Year Ending September 30, 2006, and Other Purposes*, Conference Report, 109th Cong., 1st sess., December 18, 2005, H.Rept. 109-359 (Washington: GPO, 2005), p. 514.

[124] CRS Report R44189, *Bureau of Alcohol, Tobacco, Firearms and Explosives (ATF): FY2016 Appropriations*, by William J. Krouse.

[125] U.S. Department of Justice, Bureau of Alcohol, Tobacco, Firearms, and Explosives, "ATF's History," available at http://www.atf.gov/about/history/.

state, and local governments, and other federal agencies to investigate and reduce crime involving firearms and explosives, acts of arson and bombings, and illegal trafficking of alcohol and tobacco products.[126]

Funding Narrative

Since 2005, Congress has appropriated $20 million in supplemental appropriations for the ATF. Chapter 8 of Title I of Division B of the Department of Defense, Emergency Supplemental Appropriations to Address Hurricanes in the Gulf of Mexico, and Pandemic Influenza Act, 2006 (P.L. 109-148) provided $20 million for the ATF's salaries and expenses account "to support operational recovery from hurricane-related damage in the Gulf Coast region."[127]

Federal Prison System (Bureau of Prisons)

Program Authority

Subtitle A of Title XI of the Violence Against Women and Department of Justice Reauthorization Act of 2005 (P.L. 109-162) authorized $5.066 billion for FY2006, $5.268 billion for FY2007, $5.479 billion for FY2008, and $5.698 billion for FY2009 for the Federal Prison System account.

Program Description

The Bureau of Prisons (BOP) was established in 1930 to house federal inmates, to professionalize the prison service, and to ensure consistent and centralized administration of the federal prison system.[128]

---

[126] U.S. Department of Justice, Bureau of Alcohol, Tobacco, Firearms and Explosives, Bureau of Alcohol, Tobacco, Firearms and Explosives, "Congressional Budget Submission, Fiscal Year 2016," p. 5, available at https://edit.justice.gov/sites/default/files/jmd/pages/attachments/2015/02/02/ 26._bureau_of_alcohol_tobacco_firearms_and_explosives_atf.pdf.

[127] U.S. Congress, House, *Making Appropriations for the Department of Defense for the Fiscal Year Ending September 30, 2006, and Other Purposes*, Conference Report, 109th Cong., 1st sess., December 18, 2005, H.Rept. 109-359 (Washington: GPO, 2005), p. 515.

[128] U.S. Department of Justice, Bureau of Prisons, "About the Bureau of Prisons," available at http://www.bop.gov/ about/index.jsp.

The BOP's mission is to protect society by confining offenders in prisons and community-based facilities that are safe, humane, cost-efficient, and appropriately secure, and that provide work and other self-improvement opportunities for inmates so that they can become productive citizens after they are released.[129] BOP currently operates 118 correctional facilities across the country.[130]

Funding Narrative

Since 2005, Congress has appropriated $11 million in supplemental appropriations for the BOP. Chapter 8 of Title I of Division B of the Department of Defense, Emergency Supplemental Appropriations to Address Hurricanes in the Gulf of Mexico, and Pandemic Influenza Act, 2006 (P.L. 109-148) provided $11 million for the BOP ' s buildings and facilities account "to repair hurricane-related damage in the Gulf Coast region."[131]

Office of Justice Programs

Program Authorities

Congress has not traditionally authorized appropriations for the Office of Justice Programs (OJP); rather it has authorized appropriations for grant programs administered by the OJP. The funding appropriated by Congress for the OJP under the Department of Defense, Emergency Supplemental Appropriations to Address Hurricanes in the Gulf of Mexico, and Pandemic Influenza Act, 2006 (P.L. 109-148) was not appropriated pursuant to any authorized grant program. Congress appropriated funding for OJP's State and Local Law Enforcement assistance account for the OJP to award to states affected by hurricanes in the Gulf of Mexico in 2005.

---

[129] U.S. Department of Justice, Bureau of Prisons, "Mission and Vision of the Bureau of Prisons," available at https://www.bop.gov/about/agency/agency_pillars.jsp.

[130] U.S. Department of Justice, Bureau of Prisons, "About the Bureau of Prisons," available at http://www.bop.gov/ about/index.jsp.

[131] U.S. Congress, House, *Making Appropriations for the Department of Defense for the Fiscal Year Ending September 30, 2006, and Other Purposes*, Conference Report, 109th Cong., 1st sess., December 18, 2005, H.Rept. 109-359 (Washington: GPO, 2005), p. 515.

The funding appropriated by Congress for the OJP under the U.S. Troop Readiness, Veterans' Care, Katrina Recovery, and Iraq Accountability Appropriations Act, 2007 (P.L. 110-28) was appropriated pursuant to an authorization for the Byrne Discretionary Grant program. This program was previously authorized under Part B of Subchapter V of Chapter 46 of Title 42 of the U.S. Code. However, the authorization was repealed by Section 1111(b)(1) of the Violence Against Women and Department of Justice Reauthorization Act of 2005 (P.L. 109-162). Congress continued to appropriate funding for the Byrne Discretionary Grant program until FY2011 when the program's funding was eliminated due to the earmark ban put in place by the 112th Congress.

Program Description

The OJP manages and coordinates the National Institute of Justice (NIJ), Bureau of Justice Statistics (BJS), Office of Juvenile Justice and Delinquency Prevention (OJJDP), Office of Victims of Crime (OVC), Bureau of Justice Assistance (BJA), and related grant programs. Through its component offices and bureaus, OJP disseminates knowledge and practices across America and provides grants for the implementation of crime fighting strategies. NIJ focuses on research, development, and evaluation of crime control and justice issues. NIJ funds research, development, and technology assistance, as well as assesses programs, policies, and technologies. BJS collects, analyzes, publishes, and disseminates information on crime, criminal offenders, crime victims, and criminal justice operations. BJS also provides financial and technical support to state, local, and tribal governments to improve their statistical capabilities and the quality and the utility of their criminal history records. OJJDP assists local community endeavors to effectively avert and react to juvenile delinquency and victimization. OJJDP seeks to improve the juvenile justice system and its policies so that the public is better protected, youth and their families are better served, and offenders are held accountable.

OVC distributes federal funds to victim assistance programs across the country. OVC offers training programs for professionals and their agencies that specialize in helping victims. BJA provides leadership and assistance to local criminal justice programs that improve and reinforce the nation's criminal justice system. BJA's goals are to reduce and prevent crime, violence, and drug abuse and to improve the way in which the criminal justice system functions.[132]

Funding Narrative

Since 2005, Congress has appropriated $175 million for OJP for grants to assist states affected by hurricanes in the Gulf of Mexico. Chapter 8 of Title I of Division B of the Department of Defense, Emergency Supplemental Appropriations to Address Hurricanes in the Gulf of Mexico, and Pandemic Influenza Act, 2006 (P.L. 109-148) included $125 million for OJP's State and Local Law Enforcement Assistance account for "necessary expenses related to the direct or indirect consequences of hurricanes in the Gulf of Mexico in calendar year 2005." Chapter 2 of Title IV of the U.S. Troop Readiness, Veterans' Care, Katrina Recovery, and Iraq Accountability Appropriations Act, 2007 (P.L. 110-28) included $50 million under OJP's State and Local Law Enforcement Assistance Account for the Byrne Discretionary Grant program. Language in the law stated that funds provided under this program were to be used for local law enforcement initiatives in the Gulf Coast region related to the aftermath of Hurricane Katrina. Congress also required OJP to award the $50 million it received under the U.S. Troop Readiness, Veterans' Care, Katrina Recovery, and Iraq Accountability Appropriations Act, 2007 based upon each affected state's level of reported violent crime in 2005.

[132] U.S. Department of Justice, Office of Justice Programs, Office of Justice Programs, "Congressional Budget Submission, Fiscal Year 2018," available at https://www.justice.gov/file/969001/download.

Table 22. Disaster Relief Funding for the Department of Justice (Obligations: Dollars in Thousands)

Department of Justice	Alabama	Florida	Louisiana	Mississippi	Texas	Other	Total
Criminal Division	—	—	$440	—	—	$935	$1,375
Civil Division	—	—	—	—	—	$625	$625
U.S. Attorneys	$79	$1,019	$8,806	$3,545	$627	$1,424	$15,500
U.S. Marshals Service	—	$105	$3,002	$1,066	$1,830	$2,995	$9,000
Federal Bureau of Investigation	$993	$439	$18,469	$674	$107	$24,318	$45,000
Drug Enforcement Administration	$1,906	$2	$4,302	$1,848	$135	$1,807	$10,000
Bureau of Alcohol, Tobacco, Firearms, and Explosives	—	—	—	—	—	$20,000	$20,000
Bureau of Prisons	—	$4,100	—	—	$6,900	—	$11,000
Office of Justice Programs	$26,448	—	$82,830	$65,683	$20,000	—	$195,000

Source: Unpublished data provided by U.S. Department of Justice, December 12, 2012.

Notes: The "other" categories includes funds that were not allocated specifically to an individual state, but benefited recovery efforts generally; funds that were unable to be broken out by state due to incomplete financial information (ATF only); and funds that expired. Obligations for the Office of Justice Programs includes $20 million in deobligated funds from other OJP accounts. Figures have been rounded.

Department of Labor[133]

Workforce Innovation and Opportunity Act (WIOA) Dislocated Worker Activities[134]

National Dislocated Worker Grants

The Employment and Training Administration (ETA) of the Department of Labor administers "federal government job training and worker dislocation programs, federal grants to states for public employment service programs, and unemployment insurance benefits. These services are primarily provided through state and local workforce development systems."[135]

The Workforce Innovation and Opportunity Act (WIOA, P.L. 113-128), whose programs are administered primarily by ETA, is the primary federal employment and training legislation. WIOA authorizes several job training programs: state formula grants for Adult, Youth, and Dislocated Worker Employment and Training Activities; Job Corps; and other national programs, including Native American Programs, Migrant and Seasonal Farmworker Programs, and a series of competitive grant programs authorized under Section 169 of WIOA.

ETA provides funding assistance for disaster relief activities primarily through the Dislocated Worker program, specifically by National Dislocated Worker Grants (DWG). DWGs are authorized under WIOA Section 170 and are for employment and training assistance to workers affected by major economic dislocations, such as plant closures, mass layoffs, or natural disasters.

---

133 This section was authored by David Bradley, Specialist in Labor Economics, Domestic Social Policy Division.

134 The grants described in this section are authorized by WIOA, which replaced the Workforce Investment Act (WIA; P.L. 105-220) in 2014. WIA authorized similar grants—National Emergency Grants (NEGs)—that provided the authority for grants described in Table 26, which were issued prior to WIOA. NEGs and DWGs are substantively similar.

135 U.S. Department of Labor, Employment and Training Administration, "About ETA", available at http://www.doleta.gov/etainfo/.

These DWGs are awarded primarily to states and local Workforce Development Boards (WDBs) to provide services for eligible individuals, including dislocated workers, civilian employees of the Departments of Defense or Energy employed at an installation that is being closed within 24 months of eligibility determination, employees or contractors with the Department of Defense at risk of dislocation due to reduced defense expenditures, or certain other members of the Armed Forces. Services include job search assistance and training for eligible workers. In addition, DWG funding may be used to provide direct employment ("disaster relief employment") to individuals for a period of up to 12 months for work related to a disaster.

A majority of WIOA funding for the Dislocated Worker program is allocated by formula grants to states (which in turn allocate funds to local entities) to provide training and related services to individuals who have lost their jobs and are unlikely to return to those jobs or similar jobs in the same industry. The remainder of the appropriation is reserved by DOL for a National Reserve account, which in part provides for the DWGs.[136]

Funding Narrative

The Department of Defense, Emergency Supplemental Appropriations to Address Hurricanes in the Gulf of Mexico, and Pandemic Influenza Act, 2006 (P.L. 109-148) provided $125 million in appropriations to ETA to award National Emergency Grants (NEGs) related to the consequences of hurricanes in the Gulf of Mexico in calendar year 2005. P.L. 109-148 specified that the appropriations were to remain available until June 30, 2006, and that the funds could be used to replace NEG funds previously obligated to the hurricane-impacted areas. In calendar year (CY) 2006, Alabama received $667,000, Louisiana $36.4 million, Mississippi $46.7 million, and Texas $64.9 million in NEG funding.

[136] Specifically, WIOA §132(a)(2)(A) requires that 20% of the amount appropriated for Dislocated Worker Employment and Training Activities be reserved for national dislocated worker grants, projects, and technical assistance. The remaining 80% is to be used for state formula grants.

The total of $148.6 million in NEG funding awarded to the five states, shown in Table 23, exceeds the $125 million appropriated in P.L. 109-148. In providing the award amounts and projects, ETA does not distinguish awards by funding source. Thus, some of the funding shown in Table 23 is from the NEG funding in the regular annual WIA National Reserve appropriations.[137]

The Emergency Supplemental Appropriations Act for Defense, the Global War on Terror, and Hurricane Recovery, 2006 (P.L. 109-234) provided $16 million in appropriations to ETA for "necessary expenses related to the consequences of Hurricane Katrina and other hurricanes of the 2005 season, for the construction, rehabilitation, and acquisition of Job Corps centers." P.L. 109 234 specified that the funds were to remain available until expended.

Job Corps, which is administered by ETA, is primarily a residential job training program first established in 1964 that provides educational and career services to low-income individuals ages 16 to 24, primarily through contracts administered by DOL with corporations and nonprofit organizations.

Most participants in the Job Corps program work toward attaining a high school diploma or a General Educational Development (GED) certificate, with a subset also receiving career technical training. Currently, Job Corps centers operate in 50 states, the District of Columbia, and Puerto Rico. The $16 million provided in P.L. 109-234 for construction, rehabilitation, and acquisition of Job Corps centers was most likely used for repair of the Gulfport (Mississippi) and New Orleans Job Corps centers, which were damaged during Hurricane Katrina.[138]

---

[137] NEG award amounts were obtained from ETA. ETA reports grants awarded by state and type of project in each calendar year. Because the supplemental appropriations became law December 30, 2005 (P.L. 109-148), the amounts reported in Table 23 are for calendar year 2006 only. It should be noted that additional NEG funding was provided to these five states in other calendar years. Florida, for example, received $8.5 million in NEG funding in 2005 for hurricane-related emergencies; however, given the timing of P.L. 109-148, it does not appear that Florida's funding came from the supplemental appropriations identified in Table 23.

[138] The PY (program year) 2005 Job Corps Annual Report indicated that, "Following a recent appropriation from Congress, Job Corps is on the fast track to restoring the Gulfport and New Orleans Job Corps centers, which sustained damage during Hurricane Katrina." See

Table 23. Disaster Relief Funding by the Department of Labor (Cumulative obligations as of December 2006; Dollars in Millions)

Department of Labor	Alabama	Florida	Louisiana	Mississippi	Texas	Total
Employment and Training Administration						
National Emergency Grants	$0.67	—	$36.4	$46.7	$64.9	$148.6
Office of the Secretary						
Job Corps	—	—	—	—	—	$16.0
Total	**$0.67**	**—**	**$36.4**	**$46.7**	**$64.9**	**$164.6**

Source: CRS compilation of data from the Department of Labor Employment and Training Administration, available at http://www.doleta.gov/neg/cy_awards_LastSix.cfm.

Notes: National Emergency Grant awards in were identified by reviewing the "project name" field of the Department of Labor Office of National Response data. Projects that identified Hurricanes Katrina, Rita, or Wilma were included. As noted in the text, only NEG awards for CY2006 were included in this table.

Department of Transportation[139]

DOT is the lead support agency under Emergency Support Function #1: Transportation, under the NRF. DOT reports on damage to transportation infrastructure and coordinates alternative transportation services and the restoration and recovery of the transportation infrastructure. At the time that Hurricanes Katrina, Rita, and Wilma struck, DOT also worked with FEMA in providing and coordinating transportation support, such as evacuation aid and shipping of critical supplies to the disaster area. However, by the time Gustav and Ike struck, DOT had turned over its role in evacuation aid and the shipping of critical supplies to FEMA.

During the hurricane response, DOT had only one permanent disaster program, the Federal Highway Administration Emergency Relief Program (ER). Other operating administrations, such as the Federal Aviation Administration and the Federal Transit Administration, also provided disaster assistance.

U.S. Department of Labor, *Job Corps Annual Report: Program Year July 1, 2005 - June 30, 2006*, Washington, DC, 2006, p. 27, available at http://www.jobcorps.gov/Libraries/pdf/py05report.sflb.

[139] This section was authored by Robert S. Kirk, Specialist in Transportation Policy, Resources, Science, and Industry Division.

From a budgetary perspective, however, the DOT response to the Gulf Coast hurricanes may be viewed as either DOT funding or as FEMA funding provided to DOT for the mission assignment activities assumed by its operating administrations (see Table 17, Table 18, and Table 19). Funding by the FHWA, FAA, and FTA is briefly described below, and the cumulative total allocations to the Gulf of Mexico states are provided in Table 24.

Federal Highway Administration: Emergency Relief Program (ER)

ER Program Authorities

The Federal Highway Administration's Emergency Relief Program (ER) is authorized by Title 23, U.S.C. §125 (Section 120 (e) for federal share payable).[140]

Program Description[141]

The ER program provides funds for the repair and reconstruction of roads on the federal-aid highway system that have suffered serious damage as a result of either (1) a natural disaster over a wide area, such as a flood, hurricane, tidal wave, earthquake, tornado, severe storm, or landslide; or (2) a catastrophic failure from any external cause (for example, the collapse of a bridge that is struck by a barge). Historically, however, the vast majority of ER funds have gone for natural disaster repair and reconstruction.

ER Funding for Gulf Coast Hurricane Response

ER funding allocations for Hurricanes Katrina, Rita, Wilma, Gustav, and Ike totaled almost $3.2 billion. Of this amount, just over $2.8 billion has been obligated; see Table 24.

---

140 Regulatory Reference: 23 C.F.R. Part 668.

141 See "Emergency Relief Program" available at https://www.fhwa.dot.gov/programadmin/erelief.cfm.

Funding provided for hurricane relief includes funds from the program's annual $100 million authorization and from additional sums provided in supplemental or other appropriations acts. ER funds can only be used for roads and bridges on the federal-aid highway system. Repair and reconstruction costs for other damaged roads (mostly local roads and neighborhood streets) may be reimbursed by FEMA.

Federal Aviation Administration (FAA)

FAA has approved $110.5 million for repair and improvements to hurricane-damaged airport and air traffic control infrastructure.[142] Of this amount, $40.6 million was appropriated under the Department of Defense, Emergency Supplemental Appropriations to Address Hurricanes in the Gulf of Mexico, and Pandemic Influenza Act of 2006 (P.L. 109-148). FAA also provided Airport Improvement Program discretionary funds for airport repairs in the Gulf of Mexico states.[143]

Federal Transit Administration (FTA)

The U.S. Troop Readiness Veterans' Care Katrina Recovery and Iraq Accountability Appropriations Act of 2007 (P.L. 110-28) appropriated $35 million for transit relief to the Gulf Coast states. The distribution of this funding across the Gulf Coast states is shown in Table 25. It is not unusual for FTA to be tasked by FEMA under a mission assignment to provide transit assistance to disaster victims. Table 25 does not include these FEMA-reimbursed costs.

---

[142] This total includes $1 million in Airport Improvement Program funding provided on September 19, 2008.

[143] The FAA was the lead Operational Administration for the Katrina disaster mission assignment responses. Most of the mission assignment costs overseen by FAA following Katrina were for services provided by Landstar Express America, Inc. Landstar provided transport services for the air, sea, and land transportation of supplies and resources.

Table 24. Emergency Relief Obligations for Gulf Coast Hurricanes (Obligations as of December 2012; Dollars in Thousands)

Hurricane	Alabama	Florida	Louisiana	Mississippi	Tennessee	Texas	Total
Katrina	$27,693	$29,448	$1,193,896	$1,085,905	—	—	$2,336,942
Rita	—	$793	—	—	—	$37,508	$38,301
Wilma	—	$271,462	—	—	—	—	$271,462
Gustav	—	—	$76,976	$4,825	—	—	$81,801
Ike	—	—	$17,429	—	—	$99,923	$117,352
Total	$27,693	$301,703	$1,288,301	$1,090,730	—	$137,431	$2,845,858

Source: Federal Highway Administration.

Notes: Funds are obligated through a binding agreement, such as a project agreement, entered into by the Federal Highway Administration and a state.

Table 25. Disaster Relief Funding by Modal Administration/Program (Allocated Amounts; Dollars in Thousands)

Department of Transportation	Alabama	Florida	Louisiana	Mississippi	Texas	Total
Federal Aviation Administration	$1,688	$6,356	$21,927	$73,271	$7,256	$110,498
Federal Highway Administration: Emergency Relief Program	$27,378	$523,175	$1,410,826	$1,079,712	$142,926	$3,184,017
Federal Transit Administration	$646	$475	$20,453	$12,705	$721	$35,000
Total	$29,712	$530,006	$1,453,206	$1,165,688	$150,903	$3,329,515

Source: FAA Office of Government and Industry Affairs, FTA Office of Budget, FHWA.

Notes: The FAA total includes $1 million in Airport Improvement Program funding provided for damage caused by Hurricane Ike. Totals for FAA and FTA are based on information provided to CRS as of July 13, 2010. FHWA allocations are as of January 2013. As of January 2012, FHWA began a process of identifying unobligated ER funds and withdrawing those funds no longer needed for the events for which they were allocated. Consequently, these figures could change.

Department of Veterans Affairs[144]

Medical Center in New Orleans

The Department of Veterans Affairs (VA) administers programs that provide benefits and other services to veterans and their spouses, dependents, and beneficiaries. The VA has three primary organizations to provide these benefits: the Veterans Benefits Administration (VBA), the Veterans Health Administration (VHA), and the National Cemetery Administration (NCA). The VHA provides medical care to eligible veterans and dependents. Hurricane Katrina caused extensive damage to the VA Medical Center in New Orleans.

Funding Narrative

P.L. 109-148 appropriated additional funds for necessary expenses due to the consequences of the hurricanes in the Gulf of Mexico in 2005. Funds were appropriated by category, including $198.3 million for medical services, and $26.9 million for general operating expenses, minor construction, and the National Cemetery Administration. P.L. 109-148 appropriated $367.5 million for major construction, of which $292.5 million was for a new facility in Biloxi, MS, and $75 million was for advance planning and design work to replace the VA Medical Center in New Orleans.[145]

The total amount of appropriations authorized for the new Biloxi VA Medical Center was $310 million. This amount included $292.5 million provided in. P.L. 109-148 and $17.5 million in regular appropriations. P.L. 111-212 transferred $6 million in bid savings to the Filipino Veterans Compensation Fund, and $18 million was transferred to New Orleans Medical Center project.

---

144 This section was authored by Sidath Viranga Panangala, Specialist in Veterans Policy, Domestic Social Policy Division.

145 For more information see CRS Report RL33409, *Veterans' Medical Care: FY2007 Appropriations*, by Sidath Viranga Panangala.

Later another $11 million was reprogramed from the working reserve for the new Biloxi VA Medical Center.[146] The total estimated cost of the new Biloxi VA Medical Center is $297 million. While a majority of buildings were completed in December 2011, as of FY2018 some buildings are still under construction.[147]

P.L. 109-234 appropriated $585.9 million for major construction by the VA, of which $550 million was for replacing the New Orleans Medical Center. P.L. 112-10 appropriated $310 million for FY2011, and P.L. 112-74 appropriated $60 million for FY2012, for the New Orleans Medical Center. In FY2015 $39.5 million and in FY2016 $50 million were respectively reprogrammed from the working reserve. The total estimated cost of replacing the VA Medical Center in New Orleans is approximately $1.09 billion.

The site decision for the new VA Medical Center in New Orleans was announced on November 25, 2008, and a groundbreaking ceremony was held on June 25, 2010. However, VA could not acquire all the land parcels necessary to construct the new medical center until late April 2011. The construction of the new facility began in May 2011.[148] The new medical center was formally opened on November 18, 2016,[149] and activation of various clinics would occur in various phases.[150]

---

[146] Unexpended balances that remain after construction projects are finished are placed in a working reserve and could be redirected to other major construction projects with approval from Congress or key VA officials based on the dollar threshold established in appropriation acts.

[147] Department of Veterans Affairs, *FY2018 Congressional Budget Submission Construction, Long Range Capital Plan and Appendix*, Volume 4 of 4, May 2017, p. 6-172.

[148] U.S. Congress, House Committee on Veterans' Affairs, *Deconstructing the U.S. Department of Veterans Affairs Construction Planning*, 112th Cong., 1st sess., April 5, 2011 (Washington: GPO, 2011), p.73 and p. 69.

[149] https://www.blogs.va.gov/VAntage/33256/celebrating-the-grand-opening-of-the-new-orleans-va-hospital/.

[150] Department of Veteran Affairs, Southeast Louisiana Veterans Health Care System, "Project Legacy - Frequently Asked Questions," available at https://www.neworleans.va.gov/Project_Legacy.asp.

Armed Forces Retirement Homes

Gulfport Facility

The Armed Forces Retirement Home Trust Fund provides funds to operate and maintain the Armed Forces Retirement Homes (AFRH) in Washington, DC (also known as the United States Soldiers' and Airmen's Home), and in Gulfport, MS (originally located in Philadelphia, PA, and known as the United States Naval Home). These two facilities provide long-term housing and medical care for approximately 1,600 needy veterans. The Gulfport campus, encompassing a 19 story living accommodation and medical facility tower, was severely damaged by Hurricane Katrina, and closed at the end of August 2005.

Funding Narrative

P.L. 109-148 appropriated $65.8 million for the AFRH for expenses necessary because of the Gulf of Mexico hurricanes. Of the $65.8 million, $45 million was for advance planning and design work to replace the Gulfport, MS, facility, which was nearly destroyed by Hurricane Katrina. The facility had almost 600 residents, the majority of whom were transferred to the Washington, DC, facility after the storm. P.L. 109-234 appropriated $176 million for construction of the new Gulfport facility, and consolidated an additional $64.7 million in previously appropriated funds for construction of the new facility. P.L. 110-329 and P.L. 111-117 provided additional funds ($8.0 million and $72.0 million, respectively) for construction and renovation at the Washington, DC, and Gulfport facilities (a breakdown between the facilities for the funding is not available). In October 2010, the new Gulfport facility was completed to which residents returned.

Corporation for National and Community Service[151]

The National Civilian Community Corps (NCCC), authorized under the National and Community Service Act of 1990, as amended, is a residential program for individuals age 18 through 24 that conducts projects related to, among other things, disaster preparedness and relief and recovery efforts. The $10 million in Emergency Supplemental Funds provided for NCCC in P.L. 109-234 was used to support a range of program operations in the Gulf Region, from staff and member payroll and travel to covering communications, equipment, and supply costs.[152] Funding was used in FY2007. Approximately $1.3 million went directly to the National Service Trust, which provides educational awards to NCCC members who complete 10 months of full-time service. The remaining $8.7 million was used to support program operations; it was not used to support a specific project or service. Instead, it was combined with the program's FY2007 appropriation of $26.8 million and allowed NCCC to direct members from all of its campuses to the Gulf Region for the recovery effort. The FY2007 appropriation, combined with the $8.7 million in supplemental funds, was used, among other things, to enable 1,063 members to serve 810,000 hours on 341 relief and recovery projects in the Gulf Region.

To support this work, NCCC partnered with numerous national and local organizations, local universities and churches, as well as local and federal government, including (but are not limited to) the American Red Cross; Habitat for Humanity; City Year Louisiana; The Salvation Army; Hands On Network; Federal Emergency Management Agency; St. Bernard Parish; Tulane, Xavier, and Dillard Universities; United Way of Acadiana, Louisiana; New Orleans Recovery School District; Christian Contractors Association, Mississippi; Council on Aging, Louisiana; Alliance for Affordable Energy; Arc of Greater New Orleans; Blackbelt and Central

---

[151] This section was authored by former analyst Ann Lordeman and Joselynn Fountain, Analyst in Education Policy, Domestic Social Policy Division.

[152] Information on the use of the $10 million appropriated under P.L. 109-234 was provided by the Corporation for National and Community Service in correspondence with CRS on July 15, 2009.

Alabama Housing Authority; various Boys and Girls Clubs; Mississippi Commission for Volunteers; and New Orleans Recreation Department.

Environmental Protection Agency[153]

The U.S. Environmental Protection Agency's (EPA's) primary responsibilities include the implementation of federal statutes regulating air quality, water quality, pesticides, and toxic substances; the regulation of the management and disposal of solid and hazardous wastes; and the cleanup of environmental contamination.[154] In the case of declared disasters, FEMA may call on EPA to provide assistance to state and local governments, most notably in response to releases of hazardous materials and contaminants from a major disaster or emergency.[155]

Hurricane Emergency Response Authorities

In addition to the authorities of a Presidential declaration under the Stafford Act, three federal laws authorized the development of the regulations that are embodied in the National Oil and Hazardous Substances Pollution Contingency Plan (NCP).[156] These regulations serve as EPA's standing authority and plan for response to oil spills and releases of hazardous substances.[157] Section 311 of the Clean Water Act[158] authorizes federal emergency response to oil spills into U.S. waters, onto adjoining shorelines, or that may affect natural resources under the

---

[153] This section was authored by Robert Esworthy, Environmental Specialist, Resources, Science and Industry Division.

[154] See CRS Report RL30798, *Environmental Laws: Summaries of Major Statutes Administered by the Environmental Protection Agency*, coordinated by David M. Bearden. See also U.S. EPA, "Emergency Response," available at https://www.epa.gov/emergency-response.

[155] See U.S. EPA, "Response to 2005 Hurricanes," available at https://archive.epa.gov/katrina/web/html/; and U.S. EPA, "Hurricane Sandy Response and Recovery," available at https://archive.epa.gov/region02/sandy/web/html/ indexoriginal%20response.html).

[156] See 40 C.F.R. Part 300.

[157] For further information, see CRS Report R43251, *Oil and Chemical Spills: Federal Emergency Response Framework*, by David M. Bearden and Jonathan L. Ramseur.

[158] Federal Water Pollution Control Act, P.L. 92-500, as amended, codified at 33 U.S.C. §1251 et seq.

jurisdiction of the United States.[159] The Oil Pollution Act of 1990 (OPA) amended the response authorities in Section 311 of the Clean Water Act, and established a liability and compensation framework for oil spills.[160] The Comprehensive, Environmental Response, Compensation, and Liability Act (CERCLA, commonly referred to as Superfund) authorizes federal emergency response to releases of hazardous substances into the environment.[161] The President's response authorities under these laws are delegated by executive order to the Environmental Protection Agency (EPA) in the inland zone and to the U.S. Coast Guard in the coastal zone.[162] Other response authorities apply to oil released under certain circumstances not covered by the NCP.[163]

EPA also has additional emergency response roles related to protecting water infrastructure under other response plans and authorities if required. EPA is the lead federal agency for the water sector under the National Infrastructure Protection Plan.[164] EPA also has statutory "emergency powers" under the Safe Drinking Water Act to issue orders and commence civil action if a contaminant likely to enter a public water supply system poses a substantial threat to public health, and state or local officials have not taken adequate action.[165]

---

[159] 33 U.S.C. §1321. For further discussion, see CRS Report RL33705, *Oil Spills: Background and Governance*, by Jonathan L. Ramseur.

[160] 33 U.S.C. §2701 et seq.

[161] The term "environment" includes surface and subsurface lands, surface waters, groundwater, and ambient air, making the response authorities for hazardous substances broader in terms of their physical reach than that for oil spills. 42 U.S.C. §9601 et seq. For further discussion of the authorities of CERCLA, see CRS Report R41039, *Comprehensive Environmental Response, Compensation, and Liability Act: A Summary of Superfund Cleanup Authorities and Related Provisions of the Act*, by David M. Bearden.

[162] Executive Order 12580 delegated the President's authorities under CERCLA, and Executive Order 12777 delegated the President's authorities under OPA and Section 311 of the Clean Water Act. Executive Order 13286 amended these executive orders to reflect the transfer of the U.S. Coast Guard from the Department of Transportation to the Department of Homeland Security in 2003.

[163] Subtitle I of the Solid Waste Disposal Act addresses petroleum leaked from underground storage tanks. This role is performed mainly by the states under cooperative agreements with EPA.

[164] For information on the National Infrastructure Protection Plan and sector-specific agency roles, see the Department of Homeland Security's website, available at http://www.dhs.gov/national-infrastructure-protection-plan.

[165] 42 U.S.C. §300i.

EPA Hurricane Response

EPA's primary disaster response role is carried out in accordance with the (NCP) as outlined in the NRF, Emergency Support Function 10 (ESF#10)—Oil and Hazardous Materials Annex. Under ESF#10, EPA is the lead federal agency for inland incidents and those affecting both inland and coastal zones.[166] EPA also has various other response roles under the NRF and may perform a wide array of support functions in responding to a disaster or emergency.[167] In accordance with various ESFs, EPA support to other federal agencies (primarily FEMA and the Army Corps of Engineers) and state and local governments, includes activities necessary to address threats to human health and the environment focusing on impacts to drinking water and wastewater treatment facilities and post-disaster cleanup. EPA also may support the Army Corps of Engineers in its mission under ESF #3—Public Works and Engineering Annex—to remove disaster debris[168] and cleanup of water infrastructure facilities, and to DOE under ESF #12—Energy Annex—in its effort to maintain continuous and reliable energy supplies. In practice, EPA support for this latter function has generally involved waiving environmental requirements applicable to motor vehicle fuel under the Clean Air Act. For example, as part of the federal response to hurricanes in 2005, EPA granted certain waivers under this statute in response to requests from state and local officials when significant disruptions in fuel production or distribution occurred in the wake of these natural disasters.[169]

EPA's activities following the 2005 and 2008 hurricanes included retrieval and disposal of orphan (oil) tanks and drums, the collection of household hazardous waste, and the collection of liquid and semi-liquid

---

[166] EPA is the primary agency for the inland zone and incidents affecting both inland and coastal zones; the U.S. Coast Guard has primary responsibility for coastal incidents and often acts as co-lead.

[167] For more information about EPA responsibilities under the National Response Framework, including those under individual ESFs, see EPA's "Federal Response Plans" website, available at https://www.epa.gov/emergency-response/epa-and-national-response-framework-nrf.

[168] For more information, see CRS Report R44941, *Disaster Debris Management: Requirements, Challenges, and Federal Agency Roles*, by Linda Luther.

[169] See EPA's website: "Fuels Waivers Response to 2005 Hurricanes" available at https://www.epa.gov/enforcement/ fuel-waivers#2005.

waste.[170] Additionally, EPA and Corps of Engineers staff conducted assessments, providing assistance to state and local government personnel to evaluate damages to public works. Steps involved in actually restoring service include drying out and cleaning engines; testing and repairing waterlogged electrical systems; testing for toxic chemicals that may have infiltrated pipes and plants; restoring pressure (drinking water distribution lines); activating disinfection units; restoring bacteria needed to treat wastes (wastewater plants); and cleaning, repairing, and flushing distribution and sewer lines. EPA also assisted local agencies with contaminated (non-hazardous) debris management activities.

Funding Narrative

Initially following the 2005 and 2008 hurricanes, EPA conducted assessments and provided assistance to state and local governments using existing programs and regular funding. After the initial period EPA was eligible for reimbursement by FEMA for costs associated with these efforts under a FEMA mission assignment. Funding for EPA's response to Hurricanes Katrina, Rita, Wilma, Gustav, and Ike was primarily through the FEMA mission assignments and interagency agreements with FEMA. EPA indicated that of the $505 million received cumulatively through interagency agreements for its response to the five hurricanes, $497 million was expended.[171]

In addition to the mission assignment from FEMA, EPA received a cumulative total of $21 million in emergency supplemental appropriations under P.L. 109-148 enacted December 30, 2005, and P.L. 109-234, enacted June 15, 2006. Under P.L. 109-148, EPA received $8 million in emergency

---

170 EPA activities included assessment and containment of existing Superfund sites and releases from underground storage tanks. EPA uses funds from the Superfund appropriations account to pay for emergency response activities for all pre-existing Superfund sites; see "Policy Guidance on ESF #10 Mission Assignments," available at https://www.nrt.org/sites/2/files/Suiter_Makris_Policy_Guidance_on_ESF.pdf.

171 The funding received includes $800,000 received through a U.S. Army Corp of Engineers interagency agreement. The total received amounts as reported by EPA reflected adjustments resulting from quarterly reviews on all Mission Assignments and Interagency Agreements performed jointly by FEMA/DHS, EPA Cincinnati finance office, EPA Regional Program Office, and Federal Coordinating Officers, and funding EPA provided back to FEMA.

supplemental FY2006 appropriations for the Leaking Underground Storage Tank Program (LUST) for necessary expenses to address the most immediate underground storage tank needs in areas affected by Hurricanes Katrina and Rita. P.L. 109-234 increased EPA's FY2006 appropriation by an additional $7 million for assessing underground storage tanks that may have leaked in affected areas, and made $6 million available through EPA's Environmental Programs and Management (EPM) appropriations account for increased environmental monitoring, assessment, and analytical support to protect public health during the ongoing recovery and reconstruction efforts related to the consequences of the 2005 hurricane season.

EPA provided the cumulative $15 million included in the two supplemental appropriations under the LUST program to Alabama, Louisiana, and Mississippi in the form of grants for assessment and containment of underground tanks (by statute not to exceed $85,000 per project). EPA reported no allocation of this funding to Florida or Texas. The per-state distribution was determined jointly by EPA and the affected states based on the site evaluation information available at the time. The Alabama Department of Environmental Management (ADEM) indicated completion of site work related to Katrina and initiated a return of unliquidated obligations totaling $364,670. The majority of the $6 million emergency appropriations provided within the EPA Environmental Programs and Management appropriations account was used to fund contractors for analytical and other disaster support and to purchase equipment, including replacement of expended or damaged air monitors, within Louisiana and Mississippi. Funding was also provided for similar purposes in Alabama and Florida. No EPM funding allocation was reported for Texas. EPA provided $1.4 million of the EPM supplemental funding to its Office of Research and Development and Office of Air and Radiation for continued disaster and emergency response support, including analysis in its laboratories and air monitoring, across states affected by Hurricanes Katrina and Rita.

Table 26. Disaster Relief Supplemental Appropriations for the U.S. Environmental Protection Agency (EPA): P.L. 109-148 and P.L. 109-234 (FY2006 Appropriations as Received and Distributed to States; Dollars in Thousands)

Department/Agency/Program	Alabama	Florida	Louisiana	Mississippi	Texas	Total
Environmental Protection Agency						
Leaking Underground Storage Tanks (LUST) Trust Fund	\$632	—	\$10,947	\$3,421	—	\$15,000
Environmental Programs and Management (EPM)	\$180	\$96	\$3,073	\$1,241	—	\$4,590
Total	\$812	\$96	\$14,020	\$4,662	—	\$19,590

Source: CRS interpretation of data provided by the U.S. EPA Office of the Chief Financial Officer (OCFO) through the Agency's Office of Congressional and Intergovernmental Relations (OCIR) August 2010.

Notes: P.L. 109-148 and P.L. 109-234 provided a cumulative total of \$15 million in emergency funding within EPA's LUST Trust Fund account. The Alabama Department of Environmental Management (ADEM) indicated completion of site work related to Katrina and initiated a return of unliquidated obligations totaling \$364,670. P.L. 109-234 provided \$6.0 million in emergency funding within EPA's EPM account. The total amount shown in the table for EPA's EPM account does not reflect \$1.4 million for EPA's Office of Research and Development and Office of Air and Radiation. Amounts in the table may not add due to rounding.

EPA Regular Appropriations

General appropriation funds available to states in the form of grants from EPA may also have been used in the 2005 and 2008 hurricane recovery efforts, in particular, capitalization grants from the Clean Water and the Drinking Water State Revolving Funds (SRFs).[172] The SRFs are funded within the EPA's State and Tribal Assistance Grants (STAG) appropriations account. SRF grant funding is used for local wastewater and drinking water infrastructure projects, such as construction of and modification to municipal sewage treatment plants and drinking water treatment plants, to facilitate compliance with Clean Water Act and Safe Drinking Water Act requirements, respectively.

---

[172] For more information on the Clean Water SRF, see CRS Report R44963, *Wastewater Infrastructure: Overview, Funding, and Legislative Developments*, by Jonathan L. Ramseur.

Although, following a presidentially declared emergency, public drinking water and wastewater utilities are eligible for FEMA supplemental federal disaster grant assistance for the repair, replacement, or restoration of disaster damaged facilities,[173] the portions of the annual fiscal year SRF grant allotments to states may have also been used to supplement these projects.

EPA allocates annual appropriations for these capitalization grants among the states based on an established formula authorized in the Clean Water Act and based on needs surveys under the Safe Drinking Water Act.[174] States must provide 20% matching funds in order to receive the federal funds. States combine their matching funds with the federal monies to capitalize their SRFs, which they use to issue low-interest or no interest loans to finance local water infrastructure projects in communities. The recipients generally must repay the loan to the issuing state. For FY2006-FY2011, the cumulative total allotment to the five Gulf States examined in this chapter from Clean Water SRF annual appropriations was $1.20 billion.[175] The cumulative total during the six-year period for the Drinking Water SRF was $1.16 billion.[176] What portion of these funds was used to support projects for infrastructure affected by the five hurricanes is not known.

---

[173] See U.S. EPA publication "Public Assistance for Water and Wastewater Utilities in Emergencies and Disasters," EPA 817-F-10-009, Office of Water, August 2010, available at http://water.epa.gov/infrastructure/watersecurity/emerplan/upload/Public-Assistance-for-Water-and-Wastewater-Utilities-in-Emergencies-and-Disasters.pdf. See also FEMA 322 Public Assistance Guide, under Category F, and "Federal Funding for Water/Wastewater Utilities in National Disasters (Fed FUNDS)" available at https://www.epa.gov/fedfunds.

[174] EPA must allocate the Clean Water SRF grants among the states according to a formula specified in the Clean Water Act itself, whereas the Safe Drinking Water Act authorizes EPA to develop a formula for allocating the Drinking Water SRF grants among the states that is to reflect the proportional share of each state's funding needs.

[175] Includes $438.4 million allotted to these states in FY2009 from the total $4.0 billion in CWSRF supplemental appropriations included in the American Recovery and Reinvestment Act of 2009 (ARRA, P.L. 111-5).

[176] Includes $315.4 million allotted to these states in FY2009 from the total $2.0 billion DWSRF supplemental appropriations included in the ARRA (P.L. 111-5).

The Federal Judiciary[177]

The mission of the federal courts is to protect the rights and liberties guaranteed under the Constitution. The courts are charged with interpreting and applying the law to resolve disputes through fair and impartial judgments, and ensuring fairness and equal justice for all citizens of the United States.[178]

According to the Budget Office of the Administrative Office of the U.S. Courts, Congress appropriated $18 million in emergency judiciary funding[179] for disaster relief in the aftermath of Hurricanes Katrina and Rita. These monies were obligated to (1) reimburse per diem for judges, court staff, and federal public defenders' staff who were on temporary duty assignment, and their dependents; (2) reimburse all judges and court staff who were on temporary duty assignment for travel purposes; (3) pay for furniture, equipment, and security in the temporary locations; and (4) replace furniture and equipment in courts affected by the hurricanes. Table 27 presents the funding provided to Louisiana, Mississippi, Texas, and Florida, as well as the additional funding to the Fifth Circuit.

Table 27. Disaster Relief Funding by the Federal Judiciary (Obligations current as of November 2012; Dollars in Thousands)

	Louisiana	Mississippi	Texas	Florida	Fifth Circuit	Total
The Federal Judiciary	$4,712	$881	$170	$345	$11,891	$17,999

Source: Unpublished data from the Administrative Office Budget Division.

Notes: All figures have been rounded. The Fifth Circuit encompasses the District of the Canal Zone, Louisiana, Mississippi, and Texas. The table excludes $1,360 (nominal dollars) provided to the Fourth Circuit. The Fourth Circuit encompasses Maryland, North Carolina, South Carolina, Virginia, and West Virginia.

---

[177] This section was authored by Matthew Eric Glassman, former Analyst in American National Government, and updated by Barry McMillion, Analyst in American National Government.

[178] The 94 U.S. judicial districts are organized into 12 regional circuits, each of which has a United States court of appeals. A court of appeals hears appeals from the district courts located within its circuit, as well as appeals from decisions of federal administrative agencies.

[179] P.L. 109-148, Department of Defense, Emergency Supplemental Appropriations to Address Hurricanes in the Gulf of Mexico, and Pandemic Influenza Act, 2006.

Small Business Administration[180]

Disaster Assistance Program

Authority

The Small Business Administration's (SBA) Disaster Assistance Program is authorized by the Small Business Act (P.L. 85-536, Section 7(b) 72 Stat. 387, as amended).

Program Description

The SBA's Disaster Assistance Program provides low-interest disaster loans to homeowners, renters, and businesses, as well as to private and non-profit organizations to repair or replace real estate, personal property, machinery and equipment, inventory, and business assets that have been damaged or destroyed in a declared disaster.[181]

Table 28. Small Business Administration: Number of Approved Disaster Assistance Loans for the Five Hurricanes (Number of Total Approved Applications as of January 29, 2013)

Small Business Administration	Alabama	Florida	Louisiana	Mississippi	Texas	Total
Home Loans	2,497	14,021	86,206	31,243	15,935	149,902
Business Loans	360	2,578	12,921	4,388	2,545	22,792
Economic Injury Disaster Loans	82	812	1,801	335	410	3,440
Total	2,939	17,411	100,928	35,966	18,890	176,134

Source: U.S. Small Business Administration, Office of Congressional and Legislative Affairs, correspondence with CRS on January 29, 2013.

Notes: The SBA provided disaster loans to Alabama for Hurricanes Katrina, Rita, and Wilma. Alabama did not receive loans for Hurricanes Gustav and Ike. The SBA provided disaster loans to Florida for Hurricanes Katrina, Rita, and Wilma. Florida did not receive loans for Hurricanes Gustav and Ike. The SBA provided disaster loans to Mississippi for Hurricanes Katrina, Rita, Wilma, and Gustav. Mississippi did not receive loans for Hurricane Ike. The SBA provided disaster loans to Texas for Hurricanes Katrina, Rita, Wilma, and Ike. Texas did not receive loans for Hurricane Gustav.

---

180 This section was authored by Bruce R. Lindsay, Analyst in American National Government, Government and Finance Division.

181 U.S. Small Business Administration, "Disaster Assistance," available at https://www.sba.gov/ funding-programs/ disaster-assistance.

Table 29. Small Business Administration: Approved Disaster Loan Applications by Amount (Cumulative Loan Amounts by State as of January 29, 2013; Dollars in Thousands)

Small Business Administration	Alabama	Florida	Louisiana	Mississippi	Texas	Total
Home Loans	$96,244	$450,170	$5,445,887	$2,069,160	$686,533	$8,747,994
Business Loans	$47,052	$412,085	$1,526,241	$546,417	$324,016	$2,855,811
Economic Injury Disaster Loans	$7,221	$48,917	$111,486	$19,267	$24,277	$211,167
Total	$150,517	$911,172	$7,083,615	$2,634,844	$1,034,826	$11,814,973

Source: U.S. Small Business Administration, Office of Congressional and Legislative Affairs, correspondence with CRS on January 29, 2013.

The SBA provides three categories of loans: (1) home loans, (2) business loans, and (3) Economic Injury Disaster Loans (EIDLs). Home disaster loans help homeowners and renters repair or replace disaster-related damages to homes or personal property. SBA regulations limit home loans to $200,000 for the repair or replacement of real estate and $40,000 to repair or replace personal property. Business disaster loans help business owners repair or replace disaster-damaged property, including inventory and supplies. Business loans are limited to $2 million. EIDLs provide assistance to small businesses, small agricultural cooperatives, and certain private, nonprofit organizations that have suffered substantial economic injury resulting from a physical disaster or an agricultural production disaster. EIDLs are limited to $2 million.

Table 28 lists the number of approved disaster loan applications by state and by type of loan for all five hurricanes. The actual number of loans made may be somewhat lower than the number of loan applications approved, because not all approved loan applications are subsequently accepted by the borrower. Table 29 lists the amount of the approved loans, by state.

COST-SHARES AND PROGRAMMATIC CONSIDERATIONS: HURRICANES KATRINA, WILMA, DENNIS, AND RITA[182]

Administrative and Congressional Waivers of Cost-Shares

P.L. 110-28, the "U.S. Troops Readiness, Veterans Care, Katrina Recovery, and Iraq Accountability Appropriations Act, 2007," which provided supplemental appropriations legislation for the war in Iraq and disaster recovery from Hurricane Katrina, provided cost-share reductions for disaster assistance provided to the affected states along the Gulf Coast.[183] The reductions provided to Alabama, Florida, Louisiana, Mississippi, and Texas were among the largest ever granted.

P.L. 110-28 provided a waiver of all state and local cost-shares for four disaster assistance programs that are a part of the Stafford Act. These programs included Section 403 (Essential Assistance), Section 406 (Repair, Restoration, and Replacement of Damaged Facilities), Section 407 (Debris Removal), and Section 408 (Federal Assistance to Individuals and Households). These programs are generally cost-shared in statute at 75% federal and 25% state and local with the possibility, under specified circumstances, for a 90% federal, 10% state and local ratio. Also significant was the cost-share waiver for the Other Needs Assistance Program under Section 408, which had never been waived previously. That section of Stafford states that the "Federal share shall be 75 percent."

Section 4501 of P.L. 110-28, also states in part, the following:

> (a) Notwithstanding any other provision of law, including any agreement, the Federal share of assistance, including direct Federal assistance, provided for the States of Louisiana, Mississippi, Florida, Alabama and Texas in connection with Hurricanes Katrina, Wilma, Dennis and Rita under sections 403, 406, 407, and 408 of the Robert T.

---

[182] This section was originally authored by former CRS analyst Francis X. McCarthy and was updated by Jared Brown, Analyst in Emergency Management and Homeland Security Policy.

[183] For more information on cost-shares see CRS Report R41101, *FEMA Disaster Cost-Shares: Evolution and Analysis*, by Jared T. Brown and Bruce R. Lindsay.

> Stafford Disaster Relief and Emergency Assistance Act (42 USC 5170b, 5172, 5173, and 5174) shall be 100 percent of the eligible costs under such sections.
>
> (b) APPLICABILITY
>
> 1) IN GENERAL—The federal share provided by subsection (a) shall apply to disaster assistance applied for before the date of enactment of this Act.
>
> (2) LIMITATION—In the case of disaster assistance provided under Section 403, 406 and 407 of the Robert T. Stafford Disaster Relief and Emergency Assistance Act, the Federal share provided by subsection (a) shall be limited to assistance provided for projects for which a "request for public assistance form" has been submitted.
>
> The statutory cost-share waivers were provided for five states. Per capita damage for Louisiana, Mississippi, and Alabama from Hurricane Katrina, and for Louisiana from Hurricane Rita, had already qualified those states for a decreased state cost-share (from 25% to 10%) through FEMA's regulatory formula based on estimated damage. Congress' inclusion of Florida and Texas may have been an effort to not separate out related damages within a devastating hurricane season.

Also, the decision to grant cost-share waivers to Florida and Texas may have been in recognition of the amount of help both states had provided to Mississippi and Louisiana, respectively, in both the provision of emergency management resources and in hosting large numbers of evacuees in the wake of the storms of 2005.

The "Limitation" in the legislation was intended to ensure that the projects receiving the waiver were ones already identified by applicants and not newly created projects, or perhaps, projects not necessarily related to the event that were attempting to capitalize on the reduced cost-share provision. The legislation states that a "request for public assistance" submitted prior to enactment of the bill (May 25, 2007) will require no cost-share. Any "requests for public assistance" not submitted prior to the enactment of the bill will be cost-shared at the 90% federal, 10% state and local cost-share for the affected states. This provision appeared to be intended to provide the more generous cost-share to those projects already selected by the state rather than projects that could be developed or submitted based on 100% federal funding.

There have been several instances when Congress chose to adjust a state's cost-share by legislation. Prior to large cost-share adjustments made to several FEMA programs as noted above, Congress also legislatively reduced cost-shares for states affected by Hurricane Rita.[184]

CONCLUDING OBSERVATIONS AND POLICY QUESTIONS[185]

This chapter demonstrates not only the significant amount of assistance the federal government provides for major disasters, but also the wide range of federal programs that are brought to bear to help individuals and communities respond and recover from major disasters, as well as prepare and mitigate against future disasters. Yet, this is only a partial picture of the amounts and types of disaster assistance that have been provided by the federal government on a yearly basis. The research focus for this chapter was on supplemental appropriations for the 2005 and 2008 Gulf Coast hurricanes. The federal government, however, also annually provides disaster assistance through regular appropriations and continuing resolutions, as well as supplemental appropriations. For example, with respect to the DRF, Congress provided roughly $42 billion in annual appropriations for FY2007 to FY2016 (see Table 30). This amount does not include what was provided in annual appropriations for other agencies, nor does it include what was provided through supplemental appropriations.

There are indications that expenditures on disaster assistance may increase. In recent years there has been an uptick in the number of declarations issued each year. For example, the average number of major disasters declared per year from 1953 to 2016 was 35.8.[186] However,

[184] P.L. 109-234, 115 Stat. 671.

[185] This section was authored by Bruce R. Lindsay, Analyst in American National Government, Government and Finance Division.

[186] Congress first authorized the President to issue major disaster declarations in 1953.

beginning in the 1990s there has been an uptick in the frequency with which major disasters are declared. During the 1990s the average number of major disaster declarations per year was 45.8, the average number from 2000 to 2009 was 57.1, and the average number from 2010 to 2016 was 58.7 (see Figure 1).

Table 30. Disaster Relief Fund Annual Appropriations FY2007-FY2016 (Dollars in Millions)

Fiscal Year	Amount
2007	$1,500
2008	$1,400
2009	$1,278
2010	$1,600
2011	$2,645
2012	$7,100
2013	$7,007
2014	$6,220
2015	$6,438
2016	$7,374
Total	$42,562

Source: CRS analysis of various Administration budget documents and appropriations statutes.

Note: Table 31 does not include rescissions or transfers unless they have been incorporated in appropriation acts.

Thus, while this chapter provides the most detailed information on federal assistance for the 2005 and 2008 Gulf Coast hurricanes, there is a need for further research on the subject of federal disaster assistance—including the assistance provided in response to disasters in other regions of the United States—to address existing gaps in funding information. This information would be useful because, arguably, congressional oversight and debates concerning disaster relief can be better informed with more accurate data and information on the amounts and types of assistance provided by the federal government to states, localities, and tribal nations.

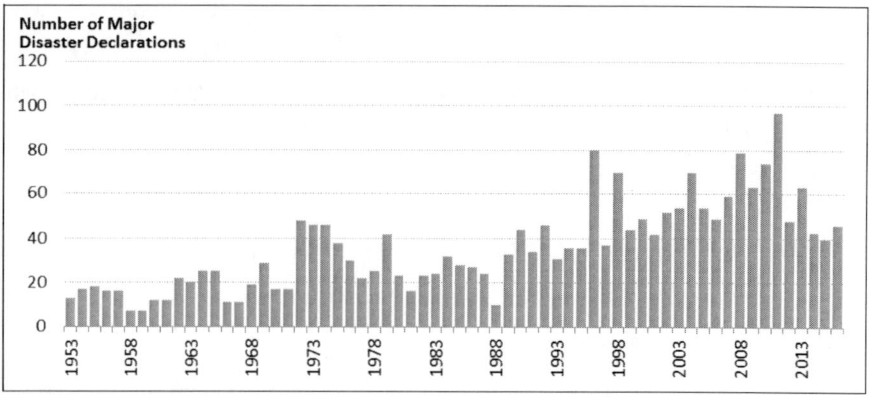

Source: CRS analysis of data derived from https://www.fema.gov/disasters and data provided by FEMA.

Figure 6. Major Disaster Declarations; 1953-2016.

Potential policy methods for addressing gaps in funding information may include requiring:

- the issuance of disaster assistance reports on an annual or quarterly basis from all federal entities that provide significant amounts of disaster assistance;
- the Office of Management and Budget (OMB) to compile a report on an annual or quarterly basis with funding information that details all federal spending for emergencies and major disasters;
- a standardization of how expenditure data are reported across federal agencies to facilitate cost comparisons;[187]
- reports to include state-specific as well as disaster-specific information. State-specific information could be used to target mitigation projects;
- disaster assistance reports to include supplemental as well as regular appropriations data;
- federal agencies to flag monies used for disaster relief that has been taken from their regular budgets; and

---

[187] Funding information is currently provided in different formats including obligations, allocations, and expenditures.

- disaster assistance reports to contain cost share information as well as detailed information on state expenditures.

Potential Methods for Controlling Costs Associated with Major Disasters

If the increase in the number of declarations and their associated costs are of concern, in addition to requiring improved data reporting Congress may choose to address the issue through a variety of policy measures.

The following sections could be used to frame a potential debate on limiting the number of declarations being issued, limiting the assistance provided after a declaration has been declared, or both.

Rationale for Keeping the Disaster Assistance the Same

To many, providing relief to disaster victims is an essential role of the government. In their view, the concern over costs is understandable given concerns over the national budget. However, they may argue that the increase in the amount of assistance provided over the past decade is justified because the occurrences of disasters are on the rise (see Figure 6).[188] The rise may be due to a number of factors including increases in inclement weather, population growth, and building development. Moreover, proponents of keeping the current system in place may say that providing assistance to disaster-stricken areas is both acceptable and needed to help a state and region's economy recover from a storm that it otherwise may not be able to recover from on its own.

Limiting the Number of Major Disaster Declarations Being Issued

Others may contend that too many major disasters are being declared and should be limited. The following sections review some policy mechanisms that could be employed to decrease the number of declarations

---

[188] For historical information on major disaster declarations. See CRS Report R42702, *Stafford Act Declarations 1953 2016: Trends, Analyses, and Implications for Congress*, by Bruce R. Lindsay.

that are being issued. The primary option consists of preventing what may be perceived by some to be marginal incidents from triggering federal assistance. Potential methods to achieve this include changing the definitions of a major disaster in Stafford Act, changing the per capita formula for determining whether a disaster is sufficiently large to warrant federal assistance, or the use of other indicators instead of, or in conjunction with, the per capita formula.

Changing the Definition of Major Disaster in the Stafford Act

Some argue that the Stafford Act has enhanced presidential declaration authority because the definition of a major disaster in Section 102(2) of Stafford Act is ill-defined.[189] Because of the expansive nature of this definition under the Stafford Act, they assert, there are not many restrictions on the types of major disasters for which the President may issue a declaration.[190] For example, some would argue that snowstorms do not warrant major disaster declarations.

Changing the Per Capita Formula

One potential method of reducing the number of major disasters being declared is to increase the per capita amount used by FEMA to make major disaster recommendations to the President. A per capita formula based on damages caused by an incident is used by FEMA to make recommendations to the President concerning whether to issue a major disaster declaration. The current per capita amount used by FEMA to make recommendations is $1.43. This amount could be increased (for example, by 10%) to reduce the number of incidents eligible for federal assistance.

If increased, Congress might require that the per capita be adjusted annually for inflation. The DHS Inspector General issued a report in May 2012, which noted that FEMA had been using a $1 per capita damage amount since 1986 for determining during its preliminary damage assessment process if it would recommend to the President that the event

---

[189] P.L. 93-288, 42 U.S.C. §5122.

[190] Richard T. Sylves, *Disaster Policy and Politics: Emergency Management and Homeland Security* (Washington, DC: CQ Press, 2008), p. 79.

was beyond the capacity of state and local governments to deal with without federal assistance. The DHS Inspector General also explained that FEMA did not begin adjusting that number for inflation until 1999. The DHS Inspector General pointed out that if the inflation adjustment had been occurring over that 13-year period, from 1986 to 1999, fully 36% fewer disasters would have qualified for a presidential declaration based on that factor.[191]

However, it is also useful to understand that the actual public announcement of factors considered for a declaration did not become public until 1999. At the behest of Congress, it was in that year that FEMA began to print the factors that were considered in regulation. Until then, all of that information had been within the "pre-decisional" part of the process in the executive branch. However, in 1999 FEMA began to identify factors considered for both Public and Individual Assistance. That is not to say FEMA was not using the per capita amount in its considerations, only that the process was not widely known or understood as it presently is. As the DHS IG notes, FEMA could have been raising that amount gradually, a process that did not begin until more than a dozen years later. On the other hand, it should also be considered that when FEMA discussed such proposals (e.g., per capita figures gradually increasing) with Congress, the result was a new Section 320 of the Stafford Act that stated:

> No geographic area shall be precluded from receiving assistance under this Act solely by virtue of an arithmetic formula or sliding scale based on income of population.

The Use of State Capacity Indicators

In 2001, the Government Accountability Office (GAO) issued a report on disaster declaration criteria. The GAO report was a comprehensive review of FEMA's declaration criteria factors. GAO recommended that FEMA "develop more objective and specific criteria to assess the

---

[191] Department of Homeland Security, Office of Inspector General, Opportunities to Improve FEMA's Public Assistance Preliminary Damage Assessment Process, pp. 5-7. Available at https://www.oig.dhs.gov/assets/Mgmt/2012/OIG_12-79_May12.pdf.

capabilities of state and local governments to respond to a disaster" and "consider replacing the per capita measure of state capacity with a more sensitive measure, such as a state's total taxable resources."

The state's Total Taxable Resources (TTR) was developed by the Department of the Treasury. GAO reported that TTR:

> is a better measure of state funding capacity in that it provides a more comprehensive measure of the resources that are potentially subject to state taxation. For example, TTR includes much of the business income that does not become part of the income flow to state residents, undistributed corporate profits, and rents and interest payments made by businesses to out-of-state stock owners. This more comprehensive indicator of state funding capacity is currently used to target federal aid to low-capacity states under the Substance Abuse and Mental Health Service Administration's block grant programs. In the case of FEMA's Public Assistance program, adjustments for TTR in setting the threshold for a disaster declaration would result in a more realistic estimate of a state's ability to respond to a disaster.[192]

It could be argued that the use of TTR would conflict with the prohibition against the use of arithmetic formulas established by Congress. However, just as FEMA's per capita measurement is one of several factors considered and not the "sole" determinant of a declaration, GAO stated that TTR would not violate Section 320 because TTR could also be used with other criteria such as those identified in regulations. Thus, some could contend that TTR could fill a similar role with perhaps more accuracy. It may also help reduce federal costs for disaster assistance by denying assistance to marginal incidents that could be otherwise handled by the state.

---

[192] U.S. Government Accountability Office, DISASTER ASSISTANCE; Improvement Needed in Disaster Declaration Criteria and Eligibility Assurance Procedures, GAO-01-837, August, 2001, p.11.

Expert Panels

Some have proposed the use of an independent expert panel to review gubernatorial requests for major disaster declarations.[193] Such panels would be comprised of individuals with specialized knowledge in certain subject areas, such as disasters, economics, and public health. The panel would take into account the severity of the incident as well as other factors that might indicate how well the state could respond to and recover from the incident. The panel would then make recommendations to the President whether the circumstances of the incident were worthy of federal assistance based on their assessment.

Some might argue that the use of an expert panel would make decisions about whether to provide assistance more objective. Others might argue that the use of a panel may slow down the declaration process and impede the provision of important assets and resources. It may be argued that the panel's recommendation would infringe on the President's authority to issue a declaration. On the other hand, it could also be argued that the President would retain the authority to issue a declaration despite the panel's recommendation.

Emergency Loans

Another potential method to reduce the number of declarations and the costs of federal disaster assistance would be to create incentives to dissuade states from requesting assistance. One method would be converting some, or all, federal assistance provided through emergency declarations into a loan program. For example, emergency declarations could be altered to provide up to a specified amount (for example, $5 billion dollars) in low interest recovery loans. Under this arrangement a state could elect to handle the incident without federal assistance rather than having to reimburse the federal government for recovery loans.

---

[193] For example, S. 1630, the Disaster Recovery Act of 2011, which was introduced on September 23, 2011, and referred to the Committee on Homeland Security and Governmental Affairs, would have amended the Stafford Act to authorize the President to declare a catastrophic incident if a recommendation was issued by an independent panel of experts.

Changes to the Stafford Act

The following section discusses some potential changes to the Stafford Act that might limit the number of declarations being issued each year.

Repeal of Section 320

As mentioned previously, Section 320 of the Stafford Act restricts the use of an arithmetic or sliding scale to provide federal assistance. Repealing Section 320 would allow formulas that establish certain thresholds that states would have to meet to qualify for assistance.

Section 404

Section 404 of the Stafford Act[194] authorizes the President to contribute up to 75% of the cost of an incident toward mitigation measures that reduce the risk of future damage, loss of life, and suffering. Section 404 could be amended to make mitigation assistance contingent on state codes being in place prior to an event. For example, states that have met certain mitigation standards could remain eligible for the 75% federal cost share. States that do not meet the standards would be eligible for a smaller share, such as 50% federal cost share. The amendment may incentivize mitigation work on the behalf of the state and possibly help reduce damages to the extent that a request for assistance is not needed, or the cost of the federal share may be lessened. The amendment could be set to take effect in three years, giving states time to act, or not.

Other Potential Amendments to the Stafford Act

Other amendments to the Stafford Act could either limit the number of declarations being issued, or the amount of assistance provided to the state by the federal government.

- The Stafford Act could be amended so that there could be no administrative adjustment of the cost-share. The cost-share could

---

[194] 42 U.S.C. §5170c.

only be adjusted through congressional action. The amendment could be designed to apply immediately.

- The Stafford Act could be amended so that federal assistance would only be available for states with corollary programs (such as Public Assistance, Individual Assistance, and housing assistance). Establishing these programs at the state level may increase state capacity to handle some incidents without federal assistance. The amendment could be designed to take effect in three years, giving states time to act, or not.
- The Stafford Act could be amended to discontinue all assistance for snow removal unless directed by Congress. The amendment could be designed to take effect in three years to give states and localities an opportunity to increase snow removal budgets, or not.

Reducing the Amount of Assistance Provided Through Declarations

Adjust the State Cost Share

Most discussions regarding state cost-shares in disaster programs and projects involve ways in which the state amount may be reduced and the federal share increased.[195] Some may contend, however, that the opposite approach should be adopted and efforts should be undertaken to reduce disaster costs by shifting the costs to the state and local level. Currently, state and local governments provide 25% of disaster costs on projects and grants to families and individuals with the federal government assuming, at a minimum, 75% of all costs.[196]

There is no statutory limit on the number of people that can be helped following a disaster.[197] Similarly, when assessing damage to state and local infrastructure there is no cap on the amount of federal funds that can be expended to make the repairs or accomplish a replacement. The only

---

[195] For additional discussion on this topic see CRS Report R41101, *FEMA Disaster Cost-Shares: Evolution and Analysis*, by Jared T. Brown and Bruce R. Lindsay.

[196] Ibid.

[197] There is however, a limit on how much any one household can receive ($33,300 at the time of this report).

limitation is that the damage must be to eligible facilities and that it is disaster-related damage.

Given that open-ended commitment by the federal government, some may argue that increasing the state share of 25% to a higher percentage would be warranted given the federal government's fiscal condition. Another option would be to make the cost-share arrangement not subject to administrative adjustment.

Disaster Loans

As mentioned previously, the assistance provided for emergency declarations could be provided through the form of loans. Similarly, some or all of the assistance provided to the state after a major disaster could be converted to low-interest or no-interest loans. For example, a state may receive the traditional 75% cost share for an incident but be required to reimburse 25% of that funding to the federal government. Loans for disaster recovery could also be incentivized. For instance, states that undertook certain pre-established preparedness mitigation measures could qualify for a larger federal share or a lower interest rate.

Policy Questions

Congress has always debated the federal role in disaster relief. In recent years the debate has intensified in light of the federal budgetary environment. Policymakers have, or may ask, a number of questions relating to federal expenditures on disaster relief to assist and improve oversight, and to better inform deliberations on legislation designed to assist individuals and communities respond and recover from incidents. Such questions may include:

- To what degree should the federal government be involved in providing disaster assistance? Is the federal government providing enough assistance, or being overly generous in providing financial assistance to states?

- Was the funding provided for the Gulf Coast storms delivered efficiently and to its intended targets? If not, how can the process be improved without slowing the provision of necessary services and resources?
- How were funding allocations to each federal entity determined? Was the process accurate, or could it be improved in upcoming disasters?
- Are there increased instances of fraud, abuse, and waste when large sums of funding are provided for disaster relief? If so, what oversight mechanisms are in place to prevent such occurrences?
- Is there unnecessary duplication of services and/or efforts given the large number of federal entities involved in disaster relief?
- The assistance provided by the federal government to the Gulf Coast was provided, in part, by a number of supplemental appropriations. Is it better to provide funding overtime through multiple supplemental appropriations, or to provide the funding once through a single supplemental appropriation?

INDEX

C

D

E

F

G

H

I

J

L

M

N

O

P

V

W

Related Nova Publications

Emergency Management: An Overview and Issues for Congress

Editor: Stephanie Padbury

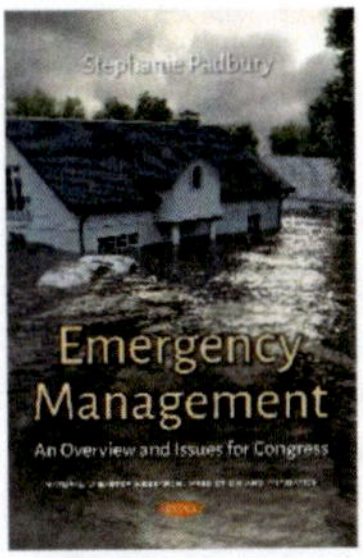

Series: Natural Disaster Research, Prediction and Mitigation

Book Description: After a flood, people are often uncertain if their eligibility for federal disaster assistance is linked to any way to whether or not they have flood insurance. The first two chapters in this book provide an overview of the assistance available to individuals and households following a flood and provides links to more comprehensive guidance on both flood insurance and disaster assistance.

Softcover ISBN: 978-1-53614-101-6
Retail Price: $82

Weathering the Storm: A State and Local Perspective on Emergency Management

Author: Anthony Rigg

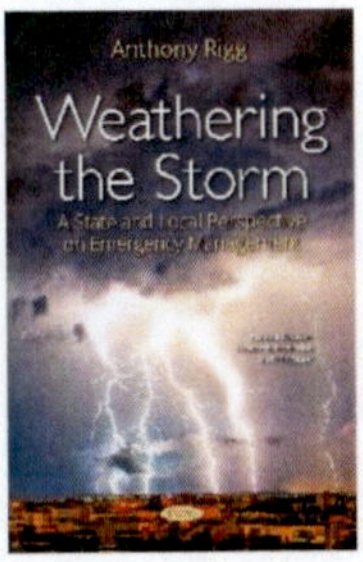

Series: Natural Disaster Research, Prediction and Mitigation

Book Description: This book is the hearing that took place in front of the U.S. House of Representatives on June 10th 2011. It provides a State and Local Perspective on Emergency Management, specifically on the efforts of State, local and non-government organizations to prepare for and respond to natural disaster, terrorist attacks and other emergencies.

Softcover ISBN: 978-1-53613-788-0
Retail Price: $79

To see a complete list of Nova publications, please visit our website at www.novapublishers.com